The Paradigm of Forests and the Survival of the Fittest

The Paradigm of Forests and the Survival of the Fittest

Editors

Sergio A. Molina-Murillo
Associate Professor/Researcher
Department of Environmental Sciences
National University of Costa Rica (UNA);
Forest Resources Unit
University of Costa Rica (UCR)
Costa Rica

Carlos Rojas
Coordinator and Researcher/Professor
Forest Resources Unit/Dpt. of Agricultural Engineering
University of Costa Rica (UCR)
San Pedro de Montes de Oca
Costa Rica

CRC Press
Taylor & Francis Group
Boca Raton London New York

CRC Press is an imprint of the
Taylor & Francis Group, an **informa** business
A SCIENCE PUBLISHERS BOOK

Cover Acknowledgement

The photograph of the spider *Nephila clavipes* was taken by Sergio A. Molina-Murillo in the low-land rainforest at Corcovado National Park in Costa Rica.

CRC Press
Taylor & Francis Group
6000 Broken Sound Parkway NW, Suite 300
Boca Raton, FL 33487-2742

First issued in paperback 2020

ISBN-13: 978-1-4987-5105-6 (hbk)
ISBN-13: 978-0-367-78320-4 (pbk)

This book contains information obtained from authentic and highly regarded sources. Reasonable efforts have been made to publish reliable data and information, but the author and publisher cannot assume responsibility for the validity of all materials or the consequences of their use. The authors and publishers have attempted to trace the copyright holders of all material reproduced in this publication and apologize to copyright holders if permission to publish in this form has not been obtained. If any copyright material has not been acknowledged please write and let us know so we may rectify in any future reprint.

Library of Congress Cataloging-in-Publication Data

Names: Molina-Murillo, Sergio A., author. | Rojas, Carlos, 1976- author.
Title: The paradigm of forests and the survival of the fittest / authors:
Sergio A. Molina-Murillo and Carlos Rojas.
Description: Boca Raton, FL : CRC Press, 2016. | Includes bibliographical
references and index.
Identifiers: LCCN 2015043737 | ISBN 9781498751056 (hardcover : alk. paper)
Subjects: LCSH: Forests and forestry--Social aspects. | Forest ecology. |
Human ecology.
Classification: LCC SD387.S55 M64 2016 | DDC 634.9--dc23
LC record available at http://lccn.loc.gov/2015043737

Visit the Taylor & Francis Web site at
http://www.taylorandfrancis.com

and the CRC Press Web site at
http://www.crcpress.com

Preface

We are all different and yet we are all here. There is simply not a single protocol to live life but the strategy around the world is still the same. Somehow, we all need each other to complement what we know about life based on our own experience. That is the way we construct knowledge but also, that is a strategy to build sustainable societies in harmony with nature. We are all fit, at least within the confines of our own reality, and so is our perception of the world and the relationship we maintain with natural resources.

This book is based on the idea that accumulated knowledge and collaboration can take our societies to healthier levels of interaction with the planet. We have always had a deep relationship with nature, but are there past lessons that we could implement in order to modify such interaction and deal with our modern environmental problems? Maybe, but once again, there is not a single solution but a series of strategies within particular contexts. However, in order to better approach these challenges, a detachment from a monothematic approach is needed. That is why this book was conceptualized with a variety of chapters. In a way, each individual chapter is a story, but put together, they portray once again our complex reality.

We hope this book serves its purpose of letting all the readers construct their own knowledge and modify their personal strategy for survival; we hope it can let us all construct a strategy for the survival of our long relationship with nature. The pages of this book contain hours of research from the different contributors around the world and reflect their good intentions for providing the best input in relation with the modern paradigm of the forests. We hope you find this book both useful and entertaining.

January 2016

Carlos Rojas
Sergio A. Molina-Murillo

Contents

Introduction

Sergio A. Molina-Murillo[1] and *Carlos Rojas*[2]

This book is intended to show in a concise manner, the social, historical and environmental framework within which humans have developed a relationship with the forests and their resources. Starting from the biological basis that permits the existence of forests to the utilization of forest resources in a modern human context, *The Paradigm of Forests and the Survival of the Fittest* is a book that has been designed to summarize the level and direction of the interaction between humans and forest ecosystems. This relationship is analyzed from different perspectives including the ecological, anthropological, economic and political aspects of environmental protection, sustainable use of natural resources, and development of human societies.

Given the characteristics of this book, rather than a monodisciplinary text, the present work encompasses the ideas of a variety of professionals from different parts of the world on the relationship between humans and forest ecosystems. As such, this book has been designed for readers from a broad range of disciplines and interests including those with affinity to subjects like environmental sciences, environmental economics, sociology, anthropology, biology, forestry and human ecology and other related disciplines. In this way, the present work aims to provoke in you, the development of an integrated attitude towards forest ecosystems and natural resources in the context of sustainability.

The book departs from a basic description of forest ecosystems and their evolution over time, which results in different forest types, along with their intrinsic characteristics. This provides the context of species

[1] Department of Environmental Sciences, National University of Costa Rica, Heredia, 30101-Costa Rica; and Forest Resources Unit, University of Costa Rica, San Pedro de Montes de Oca, 11501-Costa Rica.
E-mail: sergiomolina@una.cr
[2] Forest Resources Unit and Department of Agricultural Engineering, University of Costa Rica, San Pedro de Montes de Oca, 11501-Costa Rica.
E-mail: carlos.rojasalvarado@ucr.ac.cr

survival occurring from own competitive abilities in the framework of their relationship with human societies over time. Thus, "the survivors" are all species, and those societies that have been able to manage and protect them properly for their mutual survival. As you will find out through the passing of pages, "fittest" are those species that have been better equipped in term of their own genetic base—bio-capabilities and behavioral patterns—and in term of their contextual interactions with others. Those characteristics have translated into particular conditions for survival so the fittest have ended up being all those individuals and social groups best equipped for competition and for adaptation.

If forest ecosystems are managed and protected following "sustainable" criteria, the assumption is that there will be "progress" for human societies. Therefore, sustainable forest management practices are presented, analyzed, and discussed in several chapters. In the first one Dr. Fernandez and Dr. Markham explain how forests are distributed over a large span of latitudes and elevations, and as a response, there is large variability in their characteristics. They explore the ecological characteristics of boreal, temperate and tropical forests in terms of productivity as well as structural and species diversity. They explore the scale of variability within these ecosystems and the implications of natural and human disturbances. Forests provide a large amount of resources and environmental services for modern human populations, but disturbance caused by human activities has altered and threatened these ecosystems across the globe. In the following chapter, by Dr. Avalos, he discusses the main factors that determine forest primary production, and their implications to improve the management and conservation of ecosystems. The study of the factors that shape primary production is central to understanding key ecosystem processes, such as the ecological consequences of increased temperature and CO_2 concentration on primary production itself, the interface between the transfer of matter and energy throughout food webs, the rate of forest growth, the changes in plant succession, the relationship between diversity and ecosystem production, the expression of plant phenology, and the biological influences on climate regulation. These analyses are focused on the Tropics because it is in tropical latitudes where interactions between the biosphere and climate are more dynamic and also where the richest forests in the world are found. In such geographical area the richness of biological interactions determines subtle changes in the transfer of matter and energy which are mediated by food web processes. Tropical regions also encompass areas with extreme variation in light distribution and water availability, including systems that show fine adaptations to increase carbon gain in the face of spatial and temporal variation in water and light.

The third chapter is dedicated to the role and response of forests within climate change and modern policy frameworks. Drs. Häger and Schwendenmann highlight the challenge forest managers face today which is maximizing forest carbon storage while simultaneously ensuring other forest functions and services. They make a comprehensive review of the literature on this rapidly changing topic, and indicate that forests store between 70–80 percent of terrestrial carbon; however, on a global scale forests are lost at a rate of > 50,000 km^2/yr, and carbon fluxes from land-use change, mostly tropical deforestation, account for 8 percent of anthropogenic carbon emissions. Despite global net deforestation, forests represent 50–60 percent of the terrestrial net carbon sink, removing between 1.2–2.4 Pg C/yr from the atmosphere. The quantifications provided in this chapter are necessary to understand that forests in a modern context are extremely linked with the current ideas of "progress" and "development" and as such, it is highly unrealistic to cope with the problem of modern use of natural resources without the contextualization of global issues such as climate change.

In Chapter four Dr. Goebel provides a historical overview of the imprint of the modern world view in the state of forest ecosystems around the globe —and some of its concrete expressions such as economic systems—that have created profound environmental changes in the last several centuries. His analysis is contextualized in the overall process of the integration of forests in the economic system of national development and of global commercialization through what he calls "utilitarian conservation".

An economic view is then further expanded in chapter five by Dr. Molina-Murillo and Dr. Smith. They argue that the economic value of forest goods and services is frequently ignored in market transactions and consequently undervalued in development strategies. They discuss recent developments on the techniques for the economic valuation of forest goods and services and their monetary estimates—the value itself. This chapter is based on the premise that protecting and properly managing forestland is paramount, and economic valuation is a viable and strong option for doing this. The methodological aspects are there discussed in the context of international socio-political agendas of conservation and development, significantly influenced by the incentives societies face.

Dr. Monge brings along a discussion on an often controversial topic: forest silviculture. He discusses in chapter six the basic ecological principles governing forest silviculture, and shows a number of management examples from several regions to illustrate how forest management has been moving from a conversion approach towards a lower impact management. He argues that in order to succeed, any management regime should consider the needs and expectations of those closer or most dependent on the forest for their wellbeing (something also mentioned in chapters five and eight as well). Conversion, homogenization, rehabilitation and conservation of forest stands are all part of a dynamic forest landscape where trade-offs are unavoidable.

In chapter seven, Dr. Arevalo and Dr. Ladle bring us back to the challenges of forest conservation which in spite of having been briefly introduced in preceding chapters, are observed from the perspective of landscape ecology and scale-based strategies. Since our planet is facing unprecedented environmental change, the priorities of the global conservation movements are changing, they argue, to the creation of new protected areas, market-based mechanisms, and the sustainable use of forest resources. Since forest have multiple functions and roles, it is unsurprising that challenges of contemporary forest conservation are numerous, complex and dynamic, including conservation practices, governing models, and financial sourcing.

Dr. Sears highlights in chapter eight an over arching premise that most authors share on this book: forests are worth conserving, and there are effective and diverse ways of doing so. She reviews the linkages between forests and society, and a set of necessary conditions for the sustainable use and the conservation of forests. She provides an examination of the priorities and perspectives of distinct stakeholders and actors and the importance of forests for different sectors of society; her main focus is on the concerns of forest-dependent communities. Besides an excellent review of bibliographical sources, this chapter is written from her vast experience working in the tropics.

In summary, the reader will be delighted to explore with detail the rich discussions authors from multiple parts of the world provide here on global topics on forests ecosystems and the multiple challenges we face around them today. Authors argue that forests provide a large amount of resources and environmental services, but disturbance caused by human activities has altered and threatened them. Understanding the impacts, often as a fast progression of global forces on the primary production of forests, supports their adaptation and that of human populations. One should be hopeful of the substantial scholar advancements in recent decades to better understand forest functioning, its connection with other natural and human systems, and the development of alternative economic valuation techniques. Nevertheless, as seems to be a common concern from the authors, non-existent or conflicting legislation with complicated bureaucratic requirements, limited access to sources of education, training and empowerment, and lack of accessible sources of financing and credit to forest-dependent communities, undermine the long-term productive and conservation potential of many forest areas.

Therefore, we present this book with the hope that it will help readers develop their own paradigm of forest interaction and management in a modern context. At least we want to facilitate the discussion on the holistic importance and value of forest goods and services and on the environmental and human well-being effects of forest policies. We present a general view of the challenges surrounding the current forest paradigm as well as a magnified view of numerous managerial issues associated with it. We hope the multidisciplinary perspectives presented herein will help scholars, managers, policy-makers, and other interested actors, understand, address, and act toward sustainability.

Finally, this book is the result of many contributing friends and colleagues. We would like to thank all the participating authors for their dedication, passion, and often patience within the process of developing this book. We also like to thank Stephanie Somerville—our key assistant at the University of Costa Rica—for her detailed revision to each of the chapters in order to assure even the minor details were addressed. We are grateful with Dr. Maria Gabriela Gei and Dr. Jennifer S. Powers from the University of Minnesota and to Dr. Katherine E. Winsett from North Carolina State University, for their thoughtful revisions of earlier versions of some chapters. Last, we will like to thank the University of Costa Rica (UCR) through the Engineering Research Institute (INII) for the logistical and financial support (i.e., research projects 731-B4-900 and activity 731-A0-896), and the National University of Costa Rica (UNA) through the School of Environmental Sciences for the logistical and financial support (i.e., project 0521-13). The views expressed in these chapters are those of the various authors, and do not necessarily represent the views of their respective institutions.

CHAPTER 1

Forests in a Changing World

John Markham[1] *and Mauricio Fernández Otárola*[2,*]

ABSTRACT

Forest ecosystems have been used for human populations over millennia. Even today, forests provide a large amount of resources and environmental services for modern human populations, but disturbance caused by human activities has altered and threatened these ecosystems across the globe. Forests are distributed over a large span of latitudes and elevations, and as a response, there is large variability in their characteristics. Different kinds of forest vary in their structure, species composition, patterns of diversity, patterns of disturbance, and regeneration. Here we explore the characteristics of gymnosperm and angiosperm-dominated forests, and explore the ecological characteristics of three main forest biomes: boreal, temperate, and tropical forests in terms of productivity and structural and species diversity. Finally, we explore fine scale variability within these ecosystems and the implications of natural and human disturbance on these environments.

Introduction

Human history is intimately tied to our reliance on natural ecosystems, and our effect on them (Costanza et al. 2007). As forest cover can exist on about one third of the Earth's landmass, the structure and future

---

[1] Department of Biological Sciences, University of Manitoba, 481 Duff Roblin Building, Winnipeg Manitoba. R3T 2N2, Canada.
  E-mail: John.Markham@umanitoba.ca
[2] Escuela de Biología, Universidad de Costa Rica, 11501-2060 San Pedro, San José, Costa Rica.
* Corresponding author: maufero@gmail.com

dynamics of forests will be closely tied to human activities. Natural ecosystems have dominated the surface of the Earth for millions of years. Although it is tempting to think that humans only affected ecosystems once we became numerous and technologically advanced, we have been altering ecosystems since prehistoric times. Even prior to development of agriculture, the growth of human populations and their expansion put pressure on previously pristine ecosystems. The disappearance of the mega-fauna (giant sloths, rhinoceros, mammoths, and many other giant extinct species) at the end of the Pleistocene in the American and Asian continents was presumably one of the first large effects caused, at least partially, by human activities. The extinctions led to dramatic shifts of ecosystem structure from steppe to tundra (Zimov et al. 1995). Early human history developed under low-density populations, limited by resources and diseases. Even under these circumstances, humans can cause long lasting effects on the environment. Nomadic herding communities in Scandinavia have been shown to have effects on forest plant communities and soil 100 years after they have abandoned an area (Freschet et al. 2014) and pre-colonial activity in the Amazon created small patches of enriched soil that are still used for cultivation today (Lima et al. 2002). In other cases, pre-industrial populations' impact on the environment degraded the areas they occupied. This has repeatedly caused the collapse of whole human populations such as on Easter Island, and the Mayan cities (Diamond 2005). The invention of agriculture allowed human populations to become denser and resulted in large-scale modification of the landscape and ecosystems. The movement of human agrarian societies into forested areas obviously required large-scale clearing of land. Clearing of forests by fire not only resulted in forest removal but also changed the dominance of particular tree species in the forests that remained (Schwöer et al. 2015). A positive feedback between population growth and forest clearing developed as the more land that was cleared, the greater was the agricultural production for an area (Kimmins 1987). However, forests in these pre-industrial agrarian societies remained important sources of resources, including timber, fuel, medicines and uncultivated plant foods and wildlife. In times of conflict and famine forests were also refuges. The management of forests was therefore integral to pre-industrialized, agrarian societies (Bechmann 1990). Industrialization triggered the massive exploitation of natural resources and speeded the alteration of fairly pristine ecosystems, especially forested areas. Accompanied with this, industrialization has resulted in transport of organisms into new ecosystems, which has altered them either intentionally or unintentionally (Ward et al. 2008). Industrial agriculture and large urban populations have also resulted in eutrophication of ecosystems on a global scale (Tilman 1999) and resulted in nutrient poor communities becoming rare (Leps 2004). This, coupled with human-driven climate change, will shape the function of all ecosystems for decades to come. On the one hand, current human populations have the capacity to cause massive changes in forests in just mere decades of time compared to the thousands of years of previous civilizations. On the other hand, urbanization that resulted from the industrial revolution had also led to the return of forest cover, as lands that were previously used for agriculture are abandoned (Litvaitis 2003).

But, what is a forest? This apparently simple question is not easy to answer and a definition of a forest is quite difficult to develop to cover all ecosystems considered by some to be forests. The reason is variability; forests cover huge regions of the globe, which means, their characteristics change as a response to variation in environmental conditions. Despite all the possible variability, there is a point of agreement: there are no forests without trees—woody species over 3 m in height. But single trees do not make a forest. The underlying principle of defining forests is that groups of trees greatly modify the conditions (light, soil moisture, and chemistry) where they exist. We therefore need to consider the area covered by trees and their height and density within that area. Below a particular level of tree cover, the environment is no longer primarily affected by their presence. For example, one definition used by the FAO considers a forest to be covered by a minimum area of 0.5 ha by trees with a canopy cover of at least 10 percent (FAO 2010). These technical approaches can have important implications in management and legislation. Here we describe the characteristics of the forests, and their variability across time and space. It is impossible to cover the full diversity of forests of the world. Instead we will provide some comparisons of the major forest types, focusing on their structure and response to disturbance.

Characteristics of Forests

The main structural components of forest are trees. Competition for light drove the evolution of increased height in trees, with hydrology and structural support placing limitations on a species ultimate height (King 1990). Tree height varies widely throughout the forests of the world. Because of their height, and mass needed to maintain it, trees impose a vertical structure on the environment, making them, besides man, the ultimate ecosystems engineers. They create and maintain new conditions and substrata for the establishment and development of a full community of other organisms, greatly increasing the complexity of the environment and the dynamics of the resources (Jones et al. 1994). They also change the environmental conditions at small (internal) and large (external) scales, creating micro- and macro-environments that over evolutionary time, allow the specialization of many species. One example is the effect of trees on soil water content. During the night, when there is no light available for photosynthesis, trees in the Amazonian forest can regulate the dynamics of soil water by moving it from superficial to deeper soil layers, according to osmotic gradients, a process called hydraulic redistribution. This action functions as "water storage" when there is plenty of rain. When rain decreases, water is moved from deep to superficial soil layers during the night (hydraulic lift), following again osmotic pressures, making this vital resource available to other organisms (Oliveira et al. 2005). Also, during daily transpiration, trees move large amounts of water from the soil to the atmosphere, increasing local and large scale humidity and increasing precipitation, influencing climate over large scales (Lee et al. 2007).

Many forests show a complex vertical structure, where different strata are present. Most light is intercepted, reflected, irradiated, and filtrated by the most external layer, the canopy, and little energy reaches the lower strata. Only a small fraction (approximately two percent) reaches the bottom layer of the forest, the understory, and the ground (Montgomery and Chazdon 2001). The complexity of the intermediate area between the canopy and the understory is highly variable and depends on many factors, especially the forest height. Climatic conditions related to seasonality can modify the dynamics of light within a forest. Deciduous forests occur in both tropical and temperate regions. In the tropics they occur due to seasonally dry conditions. Although they allow the entrance of larger percentages of light during dry periods of the year, the use of this energy input can be limited by the stressed hydric conditions of the plants in these ecosystems in some areas, limiting their growth to the wet season. In temperate regions, the deciduous lifestyle of trees is due to the moderately cold, low light conditions during the winter period. Warm spring conditions allow ephemeral plants (geophytes) to grow in these regions before the trees are able to reestablish a canopy cover. A number of these have evolved unique adaptations to ensure rapid spring growth.

Forest Classification: Spatial Changes

In order to understand and manage forests, we need classification systems to sort out their diversity and predict their properties. Today, fine-scale forest type mapping is used by forest managers to predict properties such as timber yield, fire hazards, problems in stand development, and the success of planted stock. Since the late 1800s, as part of an overall plant community classification, the classification of forests developed under a very broad array of approaches. Over half a century ago the renowned plant ecologist Robert Whittaker lamented that "probably in no other field of natural science has there been such a proliferation of local schools with distinct viewpoints and techniques" (Whittaker 1962). Most classification systems were developed for specific geographic regions and ignore the properties of other regions. At the broadest scale most classification systems tend to focus on the form of the dominant vegetation rather than the species present. The most commonly recognized of these systems would be biome classification. Most classification systems also have an emphasis on the influence of climate and edaphic conditions on the distribution of vegetation. They also tend to focus on vegetation that is historically present, and on the climax community that should develop in a particular region, excluding human influences (Rowe 1972). At finer spatial scales, classification systems differ in their emphasis on which edaphic conditions to consider, successional trajectories, the presence of particular species, and subdominant vegetation. An exhaustive discussion of forest classification systems is beyond the scope of this chapter. Alternatively, as an introduction to this book we will describe two major distinction of forests, the angiosperm dominated versus gymnosperm dominated forests and the three major forest biomes, the boreal, temperate, and tropical forests.

Gymnosperm and Angiosperm Forests

Throughout the world, forests are often dominated by either gymnosperm or angiosperm trees. Gymnosperm forests tend to be found at high latitudes and altitudes, where both tree diversity and productivity is low. With a few exceptions (the larches, swamp cypresses, and Ginkgo) gymnosperm trees are evergreen, a trait common to low productivity habitats (Grime 1979). Since angiosperm trees in temperate to colder zones are mainly deciduous, angiosperm and gymnosperm forests are often inappropriately referred to as being deciduous versus coniferous, respectively. Since their evolution in the early cretaceous period, angiosperms have displaced gymnosperms from more productive habitats (Wing and Boucher 1998). A number of hypotheses have been put forward to explain the success of angiosperms. Bond (1990) has argued that the success of angiosperms in terms of their dominance over gymnosperms is a result of their more rapid reproduction, due to the evolution of the flower in the angiosperms. Angiosperms are also a much more diverse group of plants, with this diversity likely driven by their coevolution with pollinators. Along with their floral evolution, angiosperms were able to increase their photosynthetic capacity through increases in their ability to transport water (Brodribb and Field 2010). As this trait has less of an advantage as CO_2 concentrations increase, angiosperms may lose some of their dominance over gymnosperms as we add more CO_2 to the atmosphere.

It has also been argued that there are feedback loops between the presence of gymnosperm or angiosperms and soil fertility, with angiosperms creating more productive soils. One hypothesis is that gymnosperms produce leaf litter which is slow to decompose, resulting in soils with low nutrient availability, whereas angiosperms produce highly decomposable litter which increases soil nutrients (Berendse and Scheffer 2009). However, in temperate climates there are many exceptions to this pattern—in many instances soil nitrogen mineralization is higher under gymnosperm compared to angiosperm trees (Mueller et al. 2010) and in tropical climates decomposition is rapid regardless of litter quality (Davidson and Janssens 2006). Another controversial hypothesis relates to mycorrhizal control of decomposition. Trees in the Pinaceae family are the dominant gymnosperms in north hemisphere gymnosperm forests. They mainly form ectomycorrhizae while angiosperms trees mainly form arbuscular mycorrhizae (Brundrett 2002). Early studies with trenching of gymnosperm roots showed dramatic increases in decomposition suggesting ectomycorrhizae inhibit saprophytic organisms (Gadgil and Gadgil 1971). Koide and Wu (2003) suggest that a negative correlation between ectomycorrhizae and litter decomposition is mediated by the mycorrhizae's ability to decrease the water content in the soil. However, other studies have failed to show such an effect (Staaf 1988), or have shown increased decomposition associated with mycorrhizal fungal mats (Entry et al. 1991). Also, while trenching of forest plots can increase soil moisture, this can be caused by more than just changes to mycorrhizae (Fisher and Gosz 1986).

The great exception to gymnosperms being restricted to low productivity habitats is the temperate rain forest of the pacific coast of North America. These forests are dominated by gymnosperms and the species of gymnosperms that occur here tend to be the tallest species of the genera they belong to (Waring and Franklin 1979). They are all evergreen species and may attain their size and dominance due to their ability to photosynthesis in cool winter conditions compared to most of the angiosperms that are deciduous and restricted to growing in summers, where water often limits photosynthesis.

Forest Biomes

The major forest biomes (i.e., the tropical, temperate, and boreal forest) are defined by the physiognomy of the dominant plant forms. They are primarily distributed and classified by the climate in which they are found, and secondarily by local site conditions (topography and soil fertility), species composition, and age (Barbour et al. 1980). The strong relationship between climate and forest distribution is undoubtedly due to the effect of climate on plant species distribution (Zimov et al. 1995). However, the boundary between forested and non-forested biomes is often not determined by climate. In dry areas, fire, either naturally occurring or set by humans, can reduce the distribution of forests (Murphy and Bowman 2012; Butler et al. 2014). In boreal/tundra regions, aspen parklands and tropical savannas, the presence of large herbivores determines the extent of forests (McNaughton 1993; Owen-Smith 1987; Campbell et al. 1994). Human caused changes in animal abundance have therefore resulted in large-scale changes in forest distribution in many regions of the world. In these situations forest distribution is not in equilibrium with the climate. Just as with the distribution of forest boundaries, species composition may not be in equilibrium with the climate. Tree populations tend to migrate across the landscape slowly. The rate is dependent on the age of reproduction of trees, and the ability of seeds to disperse. This results in tree species migrating at different rates and the composition of different forest types change over time. We know, for instance, that after the Holocene, red cedar, one of the dominant, indicative, and culturally important species of North American temperate rainforests, did not develop into a large and abundant tree for human use until about 5000 years ago (Hebda and Mathewes 1984), whereas the other species indicative of the region were established much earlier (Brown and Hebda 1992). In New Zealand, the distribution of *Nothofagus* species is generally not found in optimal areas for their growth due to their slow rate of migration (Leathwick 1998). Similarly, trees migrate into formally glaciated regions of the northern hemisphere at species-specific rates with some species still showing a northward migration (Davis 1981). Recent, rapid, human induced climate change may result in tree species distributions that are outside of their optimum climatic niche. There is a growing consensus that we will have to move species and ecotypes of trees to new areas to accommodate this rapid climate change (Pedlar et al. 2012).

Tropical Forests

Tropical forests extend between the Tropic of Cancer in the northern hemisphere and the Tropic of Capricorn in the southern hemisphere. They are recognized by their large diversity of species and ecological interactions. Thick-stemmed woody climbers and herbaceous and woody epiphytes are important components of these ecosystems (Turner 2004). Nevertheless, there is large variability within the tropics reflected in forest characteristics. Elevational gradients cause drastic changes in forest structure and composition. In southern Central America, forests can change from the iconic tropical rain forest in the lowlands, to highland oak forest and páramo vegetation in the highest elevations, a few kilometers away.

Because of these changes, forests are classified according to their location in the elevational gradient, which is directly related to their environmental characteristics. Temperature and humidity are the main variables changing among different areas and elevations. Another component affecting the forest characteristics and classification is latitude. These main components have been used to classify the life zones of the world, providing a tool for forest's classification (Holdridge 1947). Most humid and hot forests are located in lowlands and comprise the moist forest, wet forest, and rain forest. However, within the apparently continuous cover of lowland forest areas, there are also fine- and large-scale differences. Seasonal patterns of flooding or different soil types can generate areas where specific plant species associate and even dominate. For example, in the Brazilian Amazon, rains impose seasonal patterns of flooding over large areas, and the forest is divided in várzea forest (those areas that get flooded) and terra-firme (not covered by water during the year). In the Peruvian Amazon, in the region of Loreto, edaphic associations called white-sand forests (campinarana in Brazil) are mixed within the matrix of dominant terra-firme and várzea, differing in their species' composition and structure. Another Brazilian example is the restinga forest, which covers flat areas in front of the ocean in the domain of the Atlantic forest. These restingas differ in their structure and species composition with the surrounding forest and are one of the most threatened ecosystems, mainly due to residential and touristic developments. These are examples of how a continuous forest with similar environmental characteristics can be formed by a mosaic of highly variable conditions, which generate patterns in structural changes and species associations. Nevertheless, not all tropical forests are humid, and dry forests are a main component in North, Central, and South America; the Brazilian cerrado vegetation, which ranges from a dry forest to an open savanna, originally covered an area comparable to the Amazon forest. These highly seasonal environments experience a wet season followed by dry periods causing a drastic change in the landscape and in the conditions for plants and animals. Premontane, lower montane, and montane forests dominate the middle elevations of tropical areas (approximately 400–3000 m.a.s.l.). In countries like Costa Rica, most preserved areas are categorized within this kind of forest, since the lowlands are the main environment disturbed for agriculture and human settlement purposes. Up in the mountains, above the tree line (approximately 3400 m.a.s.l.), there is the páramo, which is a kind of shrub vegetation located in the highlands of South and Central America, Costa Rica being the northern limit of distribution.

Diversity in tropical forests is extraordinary. Based in data from tropical forests plots across the world, in extreme cases more than 300 tree species wider than 10 cm in diameter at breast height (DBH) can be found in one hectare of land (average ± standard deviation: 156 ± 65; Phillips et al.

1994). The number of species would increase to several hundreds if smaller plants, terrestrial and epiphytes, were to be considered. The number of tree species is highly variable between areas, but in all cases, far larger than any temperate forest. Based on data from Phillips et al. (1994), the average number of stems wider than 10 cm DBH per hectare is 612 ± 137 ($\pm$SD), but they can reach more than 900 stems/ha.

It is usually thought, that rarity is the norm for tropical tree species, but this factor is highly variable and in many cases some species are highly abundant and are found to be dominating the forest, while some others are rare (for example densities < 1 individuals/ha). Some species can be abundant over large areas, mixed with some other less abundant ones (Condit et al. 2002). According to Ter Steege et al. (2013), of the 16000 tree species present in Amazonian forests, only 227 (1.4 percent) represent half the trees in the full ecosystem, more than 6 million-km^2, and are considered hyperdominant species. Meanwhile, another 11000 species are rare and represent only 0.12 percent of the trees in the Amazon forest. The distribution of species within the forest has been analyzed for a diverse array of species. In general, most species show aggregated patterns, but the scale of this aggregation is highly variable and is dependent on seed size, dispersal mechanisms, and regeneration strategies (Seidler and Plotkin 2006).

Forests change in time as they do in space. When a forest experiences disturbance (natural or human induced), a process of succession in the species present in this area is initiated. This successional process allows the existence of areas that differ in structure and species composition. Tropical forests are characterized by widespread, small-scale, natural disturbances that create forest gaps. Annual tree mortality is high in established forests, making them highly dynamic in terms of forest structure (Phillips et al. 1994). Individual tree mortality opens a window in the forest canopy allowing the entrance of light, which is a limiting resource in the understory. As a result of this increase of light availability, plant species begin a process of accelerated growth for acquisition of a permanent space. Some species are adapted to these disturbed environments and are called pioneer species. These species have fast growth rates and lower wood density. They also have shorter generation times in comparison to climax species which are present in mature forests. Climax species grow slowly, live longer lives, and are shade tolerant as seedlings, remaining for long periods of time in the understory. Tropical storms can also create areas of disturbance, increasing tree mortality due to strong winds and lightning. Landslides of variable scales also play an important role in steep mountainous terrains.

The existence of gaps or low scale disturbance in tropical forests increases the patterns of diversity, allowing the coexistence of pioneer and early successional species in specific areas in the mature forest. There is a mosaic of small areas under regeneration within the mature forests which theoretically should keep near a steady-state within the mature forest matrix (Chambers et al. 2013). Some areas will reach the climax vegetation while some trees will die in other points of the matrix, starting the process of regeneration. The same cannot be said of large scale human disturbance. When the scale of disturbance is larger, for example, human deforestation or a large-scale natural disturbance, large areas of forest can be altered and the process of regeneration begins. These areas will form a secondary forest. Letcher and Chazdon (2009) studied secondary forest regeneration in Costa Rican rain forests. They found no trends in stem density during regeneration, but tree biomass was higher in old secondary forests than in mature forests. By contrast, liana density decreased as regeneration advanced, but their maximum biomass was present in mature forests. The species composition of self-supporting plants increased during succession, but not for lianas. A meta-analysis of 138 studies comparing biodiversity values in mature (primary) forest, and different kinds of human disturbed areas (agricultural areas, timber plantations, pastures, selective logging, shaded plantations, etc.) showed that for all cases, disturbed areas present reduced biodiversity in comparison to primary forests (Gibson et al. 2011). This result demonstrates that primary forests are irreplaceable for biodiversity conservation in the tropics. The degree of the negative effect of disturbance on biodiversity is variable among geographic areas indicating some areas are much more susceptible than others (e.g., Asia).

Boreal Forest

The boreal region makes up 20 percent of the world's forested area. It is low in productivity and species and structural diversity (Pastor et al. 1998). The forest essentially has a two-layer structure, composed of trees and ground vegetation (Barbour et al. 1980). As with other forest types, classification within the boreal forest is often determined by edaphic gradients. Since most of the boreal area was glaciated, there are regions with very poorly developed soils and others regions with massive lacustrine deposits. Because of the low tree diversity of many forest stands, forest classification may be based on the understory plant community composition (Kimmins 1987). Cool temperatures inhibit decomposition, leading to large accumulations of organic material. As this material is accumulated, it results in nutrient immobilization and decreased plant productivity.

Unlike other forest types, stand-level disturbances characterize the boreal forest region. At high latitudes trees create long shadows. Small gaps created by single tree mortality results in little increase in light hitting the forest floor (Bonan and Shuggart 1989) and so boreal forests are less likely to be affected by gap dynamics than lower latitude forests. Fire is the predominant form of disturbance in the boreal forest. Fires are characterized by their frequency and size. Fires tend to destroy whole stands of trees and are quite variable in size. Large fires (> 10000 ha) tend to be rare events but due to their size they make up most of the burned land in the boreal forest (Stocks et al. 2002). Such large-scale fires present challenges for forest management, making predictions of local timber supply difficult. The fire cycle can be short enough to prevent any succession (Bergeron and Dubuc 1989). In the eastern North American boreal forest, fire cycles have shifted from 60 to 100 years since the late 1800s (Bergeron and Dannseau 1993). This seems to be more of a function of climate change rather than fire suppression. This increase in fire cycles has lead to an increase in the cover of coniferous species. With this greater abundance, mortality due to a conifer specific insect (spruce budworm) has increased, shifting mortality from an abiotic to a biotic effect. The multi-tree disturbances of the boreal forest can result in increased mineralization of the soil's organic layer, due to an increase in the temperature of the soil caused by increased soil radiation on the forest floor and up rooted trees lifting the soil. Human intervention in the disturbance regimes (mostly fire suppression) can therefore lead to decreased forest productivity through slowing of nutrient cycling (Baskerville 1988). Productivity can also decline with the age of the forest due to nutrients being locked up in the soil's organic layer (Wardle et al. 2002). This can result in a feedback where low soil nutrient availability leads to nutrient poor litter, which is slow to decompose (Gosz 1984). This effect is more pronounced in wet sites. On drier sites organic layer buildup may plateau in about 60 years (Ward et al. 2014).

Temperate Forest

The temperate forests are located in latitudes above the tropics and up to the zone where the arctic air mass creates conditions too cold for many species, that is where the boreal forest begins. Temperate forests also require sufficient moisture to support a tree canopy and are therefore located in regions of the globe far from mountain rain shadows (Spurr 1964). The four extensive areas of temperate forests are in East Asia, eastern North America, Europe, and New Zealand. In many ways, temperate forests are a state in between boreal and tropical forests. They have a level of tree diversity in between that of the boreal and tropical forests. Within the temperate forest there is a strong gradient of diversity from the high latitudes toward the equator. Temperate forests are taller than boreal forests and some temperate coniferous forest are the tallest in the world (Lefsky 2010). Individual tree height is much more variable in tropical forests. Consequently, the vertical structure of temperate forests is more complex that the boreal, with a shrub layer and vines, but not as complex as the tropical forest. Temperate forests also have a level of productivity that is less than tropical forests, but greater than the boreal forest. Levels of natural disturbance may also be of an intermediate scale, smaller than the massive multi-stand replacing fires of the boreal forest, but larger then single tree gaps typical of the tropical forest. In many old growth temperate stands, tree regeneration is lacking and it seems that many old stands arose from large scale disturbance events (Spurr 1964).

One of the remarkable facts of temperate forest is the variability of diversity among the large temperate forests of the world. The eastern North American temperate forest has twice as many tree species as the European temperate forest, but less than a third of the number of tree species found in the East Asian temperate forest. These differences are likely due to the geography and geological history of the regions that has hampered or aided in tree migration over time. During the last 35 million years, 70 tree species have gone extinct in Europe, compared to 20 in North America, and only five in East Asia. The east/west orientation of high mountain ranges in Europe likely made it more difficult for trees to migrate to southern refuges during the last ice age (Latham and Ricklefs 1993). East Asia being the center of evolutionary origin of many groups of woody species provides some explanation for its higher tree diversity. However, Qian and Ricklefs (1999) argue that this fact does not offer a complete explanation. A number of primitive plants that survive as relicts in Asia (most notably *Metasequoia* and *Ginkgo*) were once present in North America. This suggests that they had time to migrate to North America since their evolution, but likely did not survive due to an inability to migrate during glacial periods. East Asia differs from North America in that it has been in contact with a tropical landmass for a much longer period of time, providing plants with more opportunities to migrate as climate changes. This explanation for East Asia's greater tree diversity is supported by the fact that many of its tree groups with a high diversity have a tropical origin.

Concluding Remarks

Forests represent one of the most variable ecosystems on Earth. This factor is related to the large areas they cover at different latitudes and elevations across the globe. Their physiognomy, structure, composition, and diversity patterns change widely among different areas related to climatic patterns and historical/evolutionary factors. Forests are not homogenous in time or in space. Disturbance is a natural force influencing forest structure, composition, and diversity. Successional processes occur in mature forests as a product of natural phenomena, for example, the presence of gaps as a result of tree deaths, fires, landslides, storms, and other natural events. Human effects also create successional processes in variable scales and secondary forests are an important component in many areas. Forests gains in countries like Costa Rica during the last decades correspond to secondary forests which markedly differ from old growth (primary) forests. The length of forest successional processes varies across different environments. Despite their variability, forests represent one of the most important ecosystems for human survival. They regulate the climate, increase precipitation and soil fertility, support large species diversity, maintain large carbon stocks, and provide a large amount of resources to traditional and modern societies, including timber and non-timber forest products. They also provide environmental services, the value of which is still difficult to quantify. The alarming rate of forest disappearance during the last decades imposes a challenge for humanity to ensure forest conservation and restoration, which is linked to their own wellbeing and even survival.

References

Barbour, M.G., J.H. Burk and W.D. Pitts. 1980. Terrestrial Plant Ecology. Benjamin/Cummings Publishing Company, Menlo Park, California.

Baskerville, G.L. 1988. Redevelopment of a degrading forest system. Ambio 5: 314–322.

Bechmann, R. 1990. Trees and Man: The Forests of the Middle Ages. Paragon House, New York.

Berendse, F. and M. Scheffer. 2009. The angiosperm radiation revisited, and ecological explanation for Darwin's "abominable mystery". Ecol. Lett. 12: 865–872.

Bergeron, Y. and D. Dubuc. 1989. Succession in the southern part of the Canadian boreal forest. Vegetatio 79: 51–63.

Bergeron, Y. and P.R. Dansereau. 1993. Predicting the composition of Canadian southern boreal forest in different fire cycles. J. Veg. Sci. 4: 827–832.

Bonan, G.B. and H.H. Shugart. 1989. Environmental factors and ecological processes in boreal forests. Annu. Rev. Ecol. Syst. 20: 1–28.

Bond, W.J. 1990. Southern conifers: remnants of the deep past. Sagittarius 44: 4–7.

Brodribb, T.J. and T.S. Field. 2010. Leaf hydraulic evolution led a surge in leaf photosynthetic capacity during early angiosperm diversification. Ecol. Lett. 13: 175–183.

Brown, K.J. and R.J. Hebda. 1992. Origin, development, and dynamics of coastal temperate conifer rainforests of southern Vancouver Island, Canada. Can. J. Forest Res. 32: 353–372.

Brundrett, M.C. 2002. Coevolution of root of mycorrhizas of land plants. New Phytol. 154: 275–304.

Butler, D.W., R.J. Fensham, B.P. Murphy, S.G. Haberle, S.J. Bury and D.M.J.S. Bowman. 2014. Aborigine-managed forest, savanna and grassland: biome switching in montane eastern Australia. J. Biogeogr. 41: 1492–1505.

Campbell, C., I.D. Campbell, C.B. Blyth and J.H. McAndrews. 1994. Bison extirpation may have caused aspen expansion in western Canada. Ecography 17: 360–362.

Chambers, J.Q., R.I. Negron-Juarez, D.M. Marra, A. Di Vittorio, J. Tews, D. Roberts, G.H.P.M. Ribeiro, S.E. Trumbore and N. Higuchi. 2013. The steady-state mosaic of disturbance and succession across an old-growth Central Amazon forest landscape. P. Natl. Acad. Sci. USA 110: 3949–3954.

Condit, R., N. Pitman, E.G. Leigh, J. Chave, J. Terborgh, R.B. Foster, P. Núñez, S. Aguilar, R. Valencia, G. Villa, H.C. Muller-Landau, E. Losos and S. Hubbell. 2002. Beta-diversity in tropical forest trees. Science 295: 666–669.

Costanza, R., L. Graumlich, W.L. Steffen, C. Crumley, J. Dearing, K. Hibbard, R. Leemans, C. Redman and D.S. Schimel. 2007. What can we learn from integrating the history of humans and the rest of nature? Ambio 7: 522–527.

Davidson, E.A. and I.A. Janssens. 2006. Temperature sensitivity of soil carbon decomposition and feedbacks to climate change. Nature 440: 165–173.

Davis, M.B. 1981. Quaternary history and stability of forest communities. pp. 132–153. In: D.C. West, H.H. Shugart and D.B. Botkin (eds.). Forest Succession Concepts and Application. Springer-Verlag, New York.

Diamond, J. 2005. Collapse: How Societies Choose to Fail or Succeed. Penguin Books, New York.

Entry, J.A., C.R. Rose and K. Cromack. 1991. Litter decomposition and nutrient release in ectomycorrhizal mat soils for a Douglas fir ecosystem. Soil Biol. Biochem. 23: 285–290.

FAO. 2010. Global Forest Resources Assessment 2010. FAO, Rome.

Fisher, F.M. and J.R. Gosz. 1986. Effects of trenching on soil processes and properties in a New Mexico mixed-conifer forest. Biol. Fert. Soils 2: 35–42.

Freschet, G.T., L. Östlund, E. Kichenin and D.A. Wardle. 2014. Aboveground and belowground legacies of native Sami land use on boreal forest in northern Sweden 100 years after abandonment. Ecology 95: 963–977.

Gadgil, R.L. and P.D. Gadgil. 1971. Mycorrhiza and litter decomposition. Nature 233: 133.

Gibson, L., T.M. Lee, L.P. Koh, B.W. Brook, T.A. Gardner, J. Barlow, C. Peres, C.J.A. Bradshaw, W. Laurance, T. Lovejoy and N.S. Sodhi. 2011. Primary forests are irreplaceable for sustaining tropical biodiversity. Nature 478: 378–381.

Gosz, J.R. 1984. Biological factors influencing nutrient supply in forests. pp. 119–146. In: G.D. Bowen and E.K.S. Nambiar (eds.). Nutrition of Plantation Forests. Academic Press, London.

Grime, J.P. 1979. Plant Strategies and Vegetation Processes. John Wiley and Sons, New York.

Hebda, R.J. and R.W. Mathewes. 1984. Holocene history of cedar and native Indian cultures of the North American Pacific coast. Science 225: 711–713.

Holdridge, L.R. 1947. Determination of world plant formations from simple climatic data. Science 105: 367–368.

Jones, C.G., J.H. Lawton and M. Shachak. 1994. Organisms as ecosystem engineers. Oikos 69: 373–386.

Kimmins, J.P. 1987. Forest Ecology. Macmillan Publishing Company, New York.

King, D.A. 1990. The adaptive significance of tree height. Am. Nat. 135: 809–828.

Koide, R. and T. Wu. 2003. Ectomycorrhizas and retarded decomposition in a *Pinusresinosa* plantation. New Phytol. 158: 401–407.

Latham, R.E. and R.E. Ricklefs. 1993. Continental comparisons of temperate zone tree species diversity. pp. 294–314. In: R.E. Ricklefs and D. Schluter (eds.). Species Diversity in Ecological Communities. University of Chicago Press, Chicago.

Leathwick, J.R. 1998. Are New Zealand's *Nothofagus* species in equilibrium with their environment? J. Veg. Sci. 9: 719–732.

Lee, J.-E., R.S. Oliveira, T.E. Dawson and I. Fung. 2005. Root functioning modifies seasonal climate. P. Natl. Acad. Sci. USA 102: 17576–17581.

Lefsky, M. 2010. A global forest canopy height map from the Moderate Resolution Imaging Spectroradiometer and the Geoscience Laser Altimeter System. Geophys. Res. Lett. 37: L15401.

Leps, J. 2004. What do the biodiversity experiments tell us about consequences of plant species loss in the real world? Basic Appl. Ecol. 5: 529–534.

Letcher, S.G. and R.L. Chazdon. 2009. Lianas and self-supporting plants during tropical forest succession. Forest Ecol. Manag. 257: 2150–2156.

Lima, H.N., C.E.R. Schaefer, J.W.V. Mello, R.J. Gilkes and J.C. Ker. 2002. Pedogenesis and pre-Columbian land use of "Terra PretaAnthrosols" ("Indian black earth") of Western Amazonia. Geoderma 110: 1–17.

Litvaitis, J.A. 2003. Are pre-Columbian conditions relevant baselines for managed forests in the northeastern United States? Forest Ecol. Manag. 185: 113–126.

McNaughton, S.J. 1993. Biodiversity and stability of grazing ecosystems. pp. 361–383. *In*: E.-D. Schulze and H.A. Mooney (eds.). Biodiversity and Ecosystem Function. Springer-Verlag, Berlin.

Montgomery, R.A. and R.L. Chazdon. 2001. Forest structure, canopy architecture, and light transmittance in tropical wet forests. Ecology 82: 2707–2718.

Mueller, K.E., A.F. Diefendorg, K.H. Freeman and D.M. Eisenstat. 2010. Appraising the roles of nutrient availability, global change, and functional traits during the angiosperm rise to dominance. Ecol. Lett. 13: E1–E6.

Murphy, B.P. and D.M.J.S. Bowman. 2012. What controls the distribution of tropical forest and savanna? Ecol. Lett. 15: 748–758.

Oliveira, R.S., T.E. Dawson, S.S.O. Burgess and D.C. Nepstad. 2005. Hydraulic redistribution in three Amazonian trees. Oecologia 145: 354–363.

Owen-Smith, R.N. 1987. Pleistocene extinctions: the pivotal role of megaherbivores. Paleobiology 13: 351–362.

Pastor, J., S. Light and L. Sovell. 1998. Sustainability and resilience in boreal regions: sources and consequences of variability. Conserv. Ecol. 2: 16.

Pedlar, J.H., D.W. McKenney, I. Aubin, T. Beardmore, J. Beaulieu, L. Iverson, G.A. O'Neil, R.S. Winder and C. Ste-Marie. 2012. Placing forestry in the assisted migration debate. Bioscience 62: 835–842.

Phillips, O.L., P. Hall, A.H. Gentry, S.A. Sawyer and R. Vásquez. 1994. Dynamics and species richness of tropical rain forests. P. Natl. Acad. Sci. USA 91: 2805–2809.

Qian, H. and R.E. Ricklefs. 1999. A comparison of the taxonomic richness of vascular plants in China and the United States. Am. Nat. 154: 160–181.

Rowe, J.S. 1972. Forest regions of Canada. *In*: Fisheries and Environment Canada, Canadian Forest Service, Headquarters, Ottawa.

Schwöer, C., D. Colombaroli, P. Kaltenrieder, F. Rey and W. Tinner. 2015. Early human impact (5000–3000 BC) affects mountain forest dynamics in the Alps. J. Ecol. 103: 281–295.

Spurr, S.H. 1964. Forest Ecology. The Ronald Press Company, New York.

Seidler, T.G. and J.B. Plotkin. 2006. Seed dispersal and spatial pattern in tropical trees. PLOS Biology 4: e344.

Staaf, H. 1988. Litter decomposition in beech forests—effects of excluding tree roots. Biol. Fert. Soils 6: 302–305.

Stocks, B.J., M.S. Fosberg, T.J. Lynham, L. Mearns, B.M. Wotton, Q. Yang, J.Z. Jin, K. Lawrence, G.R. Hartley, J.A. Mason and D.W. McKenney. 1998. Climate change and forest fire potential in Russian and Canadian boreal forests. Climate Change 38: 1–13.

Ter Steege, H., N.C.A. Pitman, D. Sabatier, C. Baraloto, R.P. Salomão, J.E. Guevara et al. 2013. Hyperdominance in the Amazonian Tree Flora. Science 342: 1243092.

Tilman, D. 1999. Global environmental impacts of agricultural expansion: the need for sustainable and efficient practices. P. Natl. Acad. Sci. USA 96: 5995–6000.

Turner, I.M. 2004. The Ecology of Trees in the Tropical Rain Forest. Cambridge University Press, Cambridge, United Kingdom.

Ward, C., D. Pothier and D. Paré. 2014. Do boreal forests need fire disturbance to maintain productivity? Ecosystems 17: 1053–1067.

Ward, S.M., J.F. Gaskin and L.M. Wilson. 2008. Ecological genetics of plant invasion: what do we know? Invas. Plant Sci. Manag. 1: 98–109.

Wardle, D.A., L.R. Walker and R.D. Bardgett. 2002. Ecosystem properties and forest decline in contrasting long-term chronosequences. Science 306: 509–513.

Waring, R.H. and J.F. Franklin. 1979. Evergreen coniferous forests of the Pacific Northwest. Science 204: 1380–1386.

Whittaker, R.H. 1962. Classification of natural communities. Bot. Rev. 28: 1–239.

Wing, S.L. and L.D. Boucher. 1998. Ecological aspects of the Cretaceous flowering plant radiation. Annu. Rev. Earth Pl. Sci. 26: 379–421.

Zimov, S.A., V.I. Chuprynin, A.P. Oreshko, F.S. Chapin, J.F. Reynolds and M.C. Chapin. 1995. Steppe-tundra transition: a herbivore-driven biome shift at the end of the Pleistocene. Am. Nat. 146: 765–794.

CHAPTER 2

How Do Forests Work? Primary Production, Energy and Forest Growth

Gerardo Avalos

ABSTRACT

The amount of new biomass produced in an ecosystem per unit of time is termed primary production. The study of the factors that shape primary production is central to understanding key ecosystem processes, such as the ecological consequences of increased temperature and CO_2 concentration on primary production itself, the interface between the transfer of matter and energy through food webs, the rate of forest growth, changes in plant succession, the relationship between diversity and ecosystem production, the expression of plant phenology, and the biological influences on climate regulation. The primary objective of this chapter is to discuss the main factors that determine primary production, and their implications to improve the management and conservation of ecosystems. This analysis is focused on the Tropics because it is in tropical latitudes where interactions between the biosphere and the climate are more dynamic. In the Tropics the richness of biological interactions determines subtle changes in the transfer of matter and energy, which are mediated through food web processes. Tropical regions encompass areas with extreme variation

Escuela de Biología, Universidad de Costa Rica, 11501-2060 San Pedro, San José, Costa Rica; and The School for Field Studies, Center for Sustainable Development Studies, 100 Cummings Center, Suite 534-G, Beverly, MA 01915-6239 USA.
E-mail: gavalos@fieldstudies.org

in light distribution and water availability, including systems that show fine adaptations to increase carbon gain in the face of spatial and temporal variation in water and light. Analyzing the impacts of global warming on the primary production of tropical regions will help to answer not only basic questions in biogeochemistry but will support tropical countries to adapt to the fast progression of climate change.

Introduction

The amount of new biomass generated per unit of time is referred to as primary production. The production of new organic matter, especially new plant biomass, is mainly the result of the process of photosynthesis, and thus, of carbon sequestration and its storage in living organisms. This process is critical for ecosystem functioning, since solar energy enters living systems and food webs, first and foremost, through photosynthesis. Understanding the complexity of interactions among the factors that enhance (or limit) this process is of outmost practical importance to comprehend the dynamics of the biosphere and how it interacts with ecosystem processes, the transfer of matter and energy, and climate regulation, ultimately affecting the quality of human life.

Many factors determine the amount of new biomass that is produced in an ecosystem. In terrestrial ecosystems, water availability is a key determinant of plant growth, which affects nutrient absorption, which in turn determines the ability of plants to fix carbon. Other factors that limit primary production include temperature, light availability, as well as biological interactions (such as competition and herbivory) that determine strategies of biomass and overall resource allocation, influencing the transfer of matter and energy. In addition, the light environment and nutrient availability drive trade-offs in leaf structure and function, which in turn affect photosynthetic capacity (Nielsen et al. 1996). The concentration of greenhouse gases in the atmosphere has a direct impact on the productive capacities of ecosystems. These greenhouse gases (i.e., CO_2, methane, and water vapor) affect global temperature, and thus, intervene in the dynamics of the photosynthetic process creating negative feedback mechanisms, which at the global scale set in motion a progression that is making climate very tough for the human race. Since the start of the Industrial Revolution humankind has triggered negative feedback loops that are increasing global temperature up to 4°C (New et al. 2011). The burning of fossil fuels, global deforestation, significant changes in land cover habitat loss and cattle ranching, and the overall degradation of the biosphere have started a new climatic era (Smith and Zeder 2013) with dire consequences for humankind (Hansen et al. 2006).

Understanding the relationships between primary production and biodiversity is the objective of this chapter. Emphasis is made on tropical regions since it is in the Tropics where interactions between the biosphere and the climate are more dynamic, and the richness of biological interactions determine subtle changes in ecosystem processes with global consequences (i.e., impact of El Niño cycles on plant production) (Holmgren et al. 2001; Wright and Calderón 2006). The analysis of this information is relevant not only from a strictly scientific point of view. The exploitation of the biosphere by humanity has grown exponentially in the last 50 years. Understanding how the use of biodiversity, as well as our ecological footprint, has ramifications on the climate is of great importance for improving the management and conservation of ecosystems, and to ensure their ability to provide environmental services to the challenges of increasing population growth, biodiversity loss, and climate change.

The ability of ecosystems to sequester carbon and increase the amount of plant biomass has important implications for biogeochemical cycles, climatic variation, and changes in the availability of resources that affect the production of autotrophic organisms, mainly plants. Thus, the accurate quantification of the photosynthesis process and the monitoring of its spatial and temporal variation across different ecosystems is highly relevant to understanding the present and future responses of our planet to the process of climate change. Small economies in developing countries are not prepared to cope with climate variability associated with increasing global temperature (Adger et al. 2003). Many areas of the world suffer drastic and sudden changes in precipitation and temperature, which have a direct effect on the viability of crops and hence on food security. Tropical areas, whose population depends strictly on agricultural production, are very susceptible to rapid environmental changes combined with increasing extremes of temperature and precipitation. Droughts increased markedly in Central America and are related to an increased population density far from meeting their basic food requirements (Locatelli et al. 2011). This shows that global climate change, the production capacity of new biomass, especially food, and the degree of social welfare of a country are tightly linked.

The Process of Photosynthesis and Net Primary Production

The dynamics of the photosynthetic process can be analyzed at different scales. Biochemical processes take place instantaneously at the cellular level within the chloroplast. Light energy in the form of photons activates the chlorophyll molecule in the reaction centers of the thylakoid membranes within plant chloroplasts. Photosynthesis is thus expressed as the rate at which CO_2 is fixed per unit of area (this could be leaf area or the area of leaves per unit of ground area—i.e., leaf area index-) per unit of time. When integrated at the ecosystem level, the end products reflect the new biomass being produced, as well as the amount of CO_2 and O_2 released into the atmosphere. Carbon fluxes between the biosphere and the atmosphere can thus be estimated. Net primary production is expressed in units of energy using calories. Net primary production represents the amount of carbon given off after subtracting the carbon released as the consequence of the process of respiration. However, net primary production is usually measured as the increase in biomass without correcting for the respired carbon (Clark et al. 2001a).

The estimation of the magnitude of net primary production is crucial, as this process is closely linked to the ability of ecosystems to sequester carbon from the atmosphere and store it in different ecosystem compartments (i.e., as soil carbonates on the ocean floor or the living biomass or protoplasm), or to release it back into the atmosphere. It is necessary to improve and standardize the current methods to measure net primary production (Gower et al. 1999). However, the variety of ecosystems and the diversity of biological processes influencing the production of new biomass make this goal difficult to achieve. For example, measuring ground biomass depends on methods that are primarily destructive and highly intrusive, such as the collection of organic matter, or the monitoring of plant production, which implies the collection and measurement of whole individuals, or individual parts or modules. In other instances the manipulation of the organism is indirect but still intrusive, such as the measurement of root biomass, which affects the soil environment, thereby influencing the ability of the organism to accumulate further root biomass. Final results are extrapolated to large areas, sometimes pooling values of species for which there are no allometric equations available (Chave et al. 2005). The end result is an estimate of net primary production, sometimes a rough one.

The measurement of primary production could involve direct methods (i.e., collection of new biomass), or indirect methods (such as the generation of allometric equations to estimate whole organism biomass). Both approaches assume that all new biomass is the result of photosynthesis, with an approximate carbon content close to 47–55 percent of dry weight of plant parts. Oxygen and hydrogen compose most of the rest of the biomass, which also includes nutrients as phosphorus and nitrogen. This varies according to photosynthetic strategy (C3, C4, and CAM) as well as the growing conditions in which the plant thrives. Terrestrial ecosystems are highly dynamic, and certain processes are difficult to quantify, such as carbon loss due to root exudates, consumption of new aboveground and belowground biomass by insect herbivores, the release of volatile carbon compounds, and the natural loss of modules (roots and leaves) in between measurement intervals. Due to the latter, it is likely that the production of new biomass could be underestimated by as much as 30 percent (Clark et al. 2001a).

The Global Quantification of Primary Production Requires an Ecosystem Approach

An ecosystem consists of compartments that acquire, transform, and transfer the accumulated energy. We refer to these compartments as pools and to the transfer of energy and matter as flows or fluxes. The interface between the living things or biosphere and its interaction with physical processes constitutes an ecosystem. Light energy enters the ecosystem through photosynthesis. Energy can also enter ecosystems abiotically, through changes in temperature as well as through the effect of winds and tides that erode the rock and release materials to the atmosphere (volcanic eruptions also present a compelling example of abiotic processes that affect the energy budget of an ecosystem by increasing disturbance and releasing nutrients). Energy is lost when organic matter is oxidized back to CO_2 by combustion or through the respiration of living organisms. Energy can accumulate over long periods of time as chemical energy in coal, oil, and gas deposits.

To study the flow of matter and energy in our planet, and especially throughout the biosphere, it is necessary to implement an ecosystem approach, which facilitates the identification of the relevant components to describe the flow of matter and energy. These components are both biotic and abiotic, hence the importance of understanding the interface between biological relationships and ecosystem dynamics. Biological interactions take place in the context of community ecology. For instance, the abundance and type of flora affect the degree of carbon sequestration. The physiological status of a whole forest determines its capacity to fix and store carbon. An ecosystem approach thus provides a more holistic method to describe how our planet works, since it implies a multidisciplinary analysis where different fields of knowledge merge (i.e., thermodynamics and geological sciences).

Our planet is a very complex system. The main difficulty in understanding the flow of matter and energy lies in the multiplicity of components. Climate models have become extremely complex yet have increased predictive power. While we are limited by the collection of spatiotemporal information especially in areas of difficult access or adverse weather conditions, the existence of techniques to analyze the interaction between multivariate factors has led to the significant improvement in our ability to predict future climate scenarios.

Of the external factors that influence ecosystem processes climate is the most important one. Climatic conditions affect the rate of carbon fixation, decomposition, and respiration, among others (Holmgren et al. 2001). Differences among ecosystems can also be determined by biogeographic variation in the type of the biota, the nature of the parent material, topographic disparities, as well as the impact and nature of human activities. Time across ecological and evolutionary (geologic) scales also has important impacts, since the rate of ecosystem processes (photosynthesis, respiration, and nutrient accumulation) is time-dependent. Disturbance events, soil formation, accumulation of species diversity, and nutrient buildup take place over geologic scales. Many ecosystem processes follow a chronosequence, this is, a spatial gradient subject to different disturbance regimes that take place over time (i.e., the historical influence of variable tide levels of Amazonian rivers and how they have influenced plant succession and plant species abundance and composition (Losos 1995)). Over evolutionary time, the composition of the biota has changed across this chronosequence, affecting at the same time ecosystem processes such as biomass accumulation, carbon sequestration, and carbon fixation (Walker et al. 2010). The level of resources, and whether they become limiting for organisms, is also influenced by a chronosequence.

Net Primary Production

Net primary production (NPP) is the net carbonic gain of an ecosystem coming from the CO_2 fixed by autotrophic organisms, be these plants or autotrophic microbes. In other words, NPP corresponds to the balance between the amount of production of new biomass minus the amount of carbon given off as part of the respiration process, plus the carbon lost through combustion and the decomposition of organic matter. It is important to differentiate NPP from net ecosystem production, or NEP, a concept that refers to the accumulation of new biomass at the ecosystem level after correcting for the carbon that is lost due to plant respiration (and other autotrophic organisms), respiration of heterotrophic organisms, runoff, and carbon lost by disturbances, which include landslides and other events that expose the organic soil, as well as volcanic eruptions.

Relevance of the Photosynthetic Process

Photosynthesis is the main route through which carbon enters ecosystems, and it is the most dynamic process. Other mechanisms for carbon sequestration are no less important, but occur at very slow rates and take thousands of years to accumulate carbon in large quantities. This is the case of the carbonates stored in the deep ocean layers and the organic matter amassed as coal, oil and natural gas, and in the tundra soils. Approximately 50 percent of the carbon sequestered on our planet is fixed due to the photosynthesis that takes place in the oceans (Riebesell et al. 2007). The general trend is to increase the C:N ratio in the oceans with increasing CO_2 atmospheric concentration. The consequences of this phenomenon are not trivial since carbon over-consumption of carbon by marine organisms will increase the magnitude of ocean acidity, and will also decrease the concentration of oxygen, with significant negative effects on marine food webs through a decreased capacity to decompose organic matter. The overall process of ocean acidification will have dramatic impacts on the upper level plankton as well as on marine communities that precipitate carbonates (especially coral reefs), causing a decrease in species diversity and associated ecosystem functioning (Feely et al. 2009).

Carbon exists in the atmosphere as a trace component with a concentration of 398.55 ppm, or 0.039 percent. Current data from NOAA indicates a steady increase, with CO_2 concentration approaching 400 ppm (NOAA 2015). Human activities have augmented very rapidly this concentration compared to preindustrial levels (280 ppm). The use of fossil fuels is responsible for 75 percent of current CO_2 emissions, with deforestation, increased urban sprawling and overall changes in vegetation cover responsible for 25 percent of the emissions (ACS 2015). This carbon is mainly in the form of CO_2, which is functionally associated with the rise in temperatures and the high turnover rates or flows between the different climatic compartments.

It is estimated that approximately 762 Gt of carbon is kept in the atmosphere as CO_2 (Fig. 1). The living biomass and the carbon stored in the soil represent around 2,541 Gt. Oceans store 38,118 Gt of carbon dissolved in different forms of carbonates, mainly in the deeper layers of the ocean bottom. Carbon maintained in deep ocean carbonates has a very low turnover rate. The residence time of a carbon atom in the atmosphere is only of 5 years, whereas in the living vegetation the average is 10 years. In ocean sediments, carbon can stay chemically inactive hundreds or even millions of years, especially if it is stored as a fossil fuel. However, it is estimated that this turnover rate will increase as temperature rises, and with the increase in atmospheric turbulence our planet will see augmented

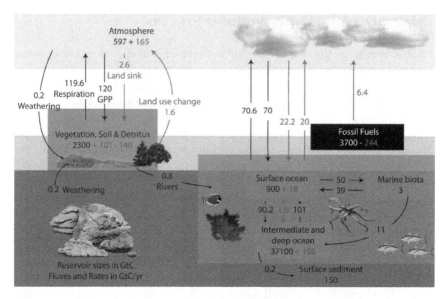

Fig. 1. Sinks, sources, and turnover rates of carbon in the global carbon cycle.

climate instability. The Earth is a classic example of a positive feedback system in which small changes in the amount of energy that is absorbed and retained have synergistic effects on atmospheric circulation, ocean currents, and thus, major climatic events. The sequestration of CO_2 from the atmosphere by living systems (the oceanic and land-based photosynthesis) removes 55 percent of carbon emissions from fossil fuel consumption and deforestation. Clearly we have a significant deficit in carbon sequestration that will not be corrected in the short term unless cleaner and more efficient sources of energy are implemented immediately. The current economy is based on the consumption of fossil fuels, and this pattern will not change in the short term.

Photosynthetic Strategies

The ancestral photosynthetic condition in terrestrial plants is the C3 photosynthetic pathway. The initial product of carbon fixation, in this case, is a 3-carbon compound, the phosphoglyceric acid (PGA). C3 plants originated during the Paleozoic and Mesozoic, and today dominate the flora, being abundant in moist environments with intermediate light conditions. In this strategy, light energy splits the water molecule producing oxygen gas, hydronium ions, and electrons. Electrons enter the electron transport chain in the membrane of the thylakoid where proteins form a proton pump that mediates ATP production. In addition, the hydrogen coming from the split water molecule reduces NADP to NADPH, which is then used in the Calvin Cycle. In this first phase, light energy is required to produce fissionable water and reduced NADPH. In the second phase, known as the dark reactions, the energy conversion into ATP and NADPH is completed. This conversion can occur in the dark or independently of light. However, many of the reactions required for the dark phase need the activation of enzymes during the light phase, so that there is no clear distinction between light-dependent and dark reactions. The dark reactions take place in the internal liquid of the thylakoid, or stroma, following the Calvin cycle. In these reactions, atmospheric CO_2 is fixed by the enzyme Rubisco (ribulose 1-5 bisphosphate carboxylase oxygenase) into a 3-carbon compound, the 3-fosfogliceraldehyde phosphate (3PGA) using the chemical energy stored in ATP and NADPH from the light-dependent reactions. The 3PGA enters the Calvin Cycle and subsequently generates glyceraldehyde-3-phosphate, which is then metabolized into glucose, follows the metabolism of sugars, and is involved in the production of starch, or could enter the lipid metabolism.

Another photosynthesis pathway, the C4 pathway, evolved at the end of the Cretaceous in response to a decrease in the concentration of CO_2 (Sage 2004). Here, the initial product of carbon fixation is a 4-carbon acid, oxaloacetate. C4 plants have not abandoned completely the C3 strategy, but instead, concentrated it within the cells of the vascular bundle sheath. These are enlarged parenchyma cells that have thicker walls and cover hermetically the vascular bundles in the leaf. Within the bundle sheath cells, chloroplasts occupy a large volume and have little or no grana at all, showing a large concentration of starch granules. In contrast, the regular mesophyll cells maintain smaller chloroplasts with almost no starch granules and with well-developed grana (Sine 2002).

This spatial separation between the mesophyll and bundle sheath cells has mechanistic consequences for carbon fixation, which determines whether a plant uses a C3 or C4 photosynthetic pathway. The mesophyll cells fix carbon into oxaloacetate using the enzyme phosphoenolpyruvate carboxylase (PEP). PEP shows poor affinity to fix oxygen under high temperatures, which is in contrast with the high affinity for oxygen of the C3 enzyme Rubisco. The affinity of Rubisco for oxygen, increases with temperature, leading to the process of photorespiration. C3 plants increase photorespiration, and thus fix oxygen instead of carbon under various stress conditions, such as lack of water, increased radiation, and increased temperatures. This leads to lower photosynthetic rates, and thus, less carbon fixation.

It is in the cells of the vascular bundle where Rubisco is highly concentrated. These cells also increase the internal CO_2 concentration. In the bundle sheath cells the Calvin Cycle proceeds normally since in the regular C4 pathway oxaloacetate is converted to malate by a decarboxylation reaction that gives off CO_2 using the highly concentrated Rubisco. Through this increased Rubisco and CO_2 concentration mechanism, C4 plants can continue photosynthesizing normally at temperatures in which C3 plants would be doing photorespiration (Sine 2002). For this reason, C4 plants tend to be more abundant in humid and warm environments, and in general, under any environmental condition that could lead to the stress of the photosynthetic process for C3 plants. However, under conditions of low temperatures, or milder temperatures, C4 plants do not have more advantages than C3 plants, since the C4 pathway tends to be more expensive due to the intermediate steps involved.

In the third photosynthetic pathway, Crassulacean Acid Metabolism (CAM), CO_2 is first carboxylated as phosphoenolpyruvate (PEP) by the PEP enzyme. The first product of this carboxylation is oxaloacetic acid (OAA), a 4-carbon compound. In contrast to C4 photosynthesis, CAM plants predominantly fix CO_2 at night when the stomata are open. CAM plants evolved under dry conditions where water availability was very limited (Keeley and Rundel 2003). In CAM plants there is an evolutionary trend towards epiphytism and the colonization of habitats almost totally independent from the ground (atmospheric environments). In CAM plants, the stomata remain closed during the day to prevent water loss, but are open at night when CO_2 is fixed first as oxaloacetate, and then as malic acid in the vacuoles of mesophyll cells by the enzyme PEP. To cope with increased pH in the mesophyll, many CAM plants are succulent. During the day, the stomata close and the malic acid is transferred from the vacuole to the mesophyll cells where it enters the regular Calvin Cycle. CAM plants, as well as C4 and C3-C4 and CAM intermediate species, have higher water use efficiency. CAM species, although not extremely diverse (approximately 7 percent of the current angiosperms are CAM) have managed to colonize a wide variety of habitats beyond atmospheric and epiphytic environments, including deserts, swamps, and seasonally flooded sites. Also, many poikilohydric plants, which can almost completely lose their whole water content, and then can recuperate their internal moisture level quickly, show CAM photosynthesis, which is particularly advantageous when water availability is unpredictable or highly seasonal. This demonstrates that CAM metabolism can be combined with desiccation tolerance, besides being a pathway that specifically favors water conservation (Lüttge 2004).

Factors that Influence the Photosynthetic Process and Plant Primary Production

Of all the environmental factors that affect plant growth, light is the most variable. Light intensity in the form of photosynthetically active radiation or PAR (400–700 nm), may vary up to three orders of magnitude within the same ecosystem. In a mature tropical rainforest, the complexity of vegetation structure creates a multilayered environment where shade-tolerant species located in the understory coexist with light-demanding species or pioneers that exploit high-light conditions. Shade-tolerant species may change their light requirements as they grow towards the canopy expressing ontogenetic niche shifts (Kitajima and Poorter 2008). However, some shade-tolerant species may remain their whole life and complete their cycle within the forest understory. Between the extremes of shade tolerance, shade avoidance, and light-demanding species there is a diverse range of regeneration strategies, expressing ontogenetic variation as they move from seedlings and juveniles to adults, escaping the light-deprived environment of the understory. The wide variation in light distribution over space and time has influenced the evolution of different life history strategies to complete the life cycle (Dietze and Clark 2008). These strategies go beyond the traditional classification based on gap-phase disturbances, and are affected by fine environmental changes, as well as by fine changes in the distribution of size classes and their influence on population dynamics across environmental gradients (see Otárola and Avalos 2014).

The light that reaches the canopy can get to levels close to 2,000 umol/ m^2/s. This light is reduced by 95 percent within the first meter below the canopy surface (Avalos et al. 2007). In the understory, under deep canopy cover, the light intensity can only reach 1 percent or less of what is received on the canopy. This extreme variation in light gradients has been proposed as one of the main drivers of niche partitioning in tropical forests, and thus, as one major cause of tropical species diversity (Brokaw and Busing 2000; Silvertown 2004).

In understory plants, which experience about 1 percent of the PAR reaching the canopy, the induction of the photosynthetic apparatus allows the efficient use of flecks of light within the forest floor (see Chazdon and Pearcy 1991). Sunfleck dynamics and their spatial and temporal variation represent a classic area of ecophysiological research on tropical plants. A sunfleck consists of a transient beam of light that reaches the forest floor with a magnitude similar to that of the light that reaches the canopy. Sunflecks are highly energetic, but their spatial distribution is unpredictable (Chazdon and Pearcy 1986). Although there is high spatial uncertainty, the temporal distribution of sunflecks is aggregated, so that understory plants can be photoinduced and progressively increase their photosynthetic rate to use subsequent sunflecks with higher efficiency, which come temporally aggregated. Similar photosynthetic dynamics has been observed in deeply shaded environments under the thick foliage of tropical lianas, down to about 40–100 cm below the canopy surface (Avalos and Mulkey 1999; Avalos et al. 2007). Most of the primary production of understory plants (and shade-adapted plants in general) is due to the efficient utilization of sunflecks (Chazdon 1988).

The dynamics of sunfleck distribution and variation shows that tropical ecosystems could efficiently use subtle changes in light. For instance, although most tropical forests have minimal variation in photoperiod, it is sufficient to stimulate morphogenetic responses, as in the expression of different phenological characteristics (Borchert 1994; Rivera and Borchert 2001; Wright 1996). Photoperiodic control of phenological traits, such as leaf fall and leaf flush, has been observed in tropical stem-succulent trees (Borchert and Rivera 2001). The synchronization of leaf fall takes place after a decrease in daylength following the autumn equinox. Flushing happens weeks before the first rains after the spring equinox when daylength increases (Borchert et al. 2005). Similar patterns have been observed in the tropical dry forest understory shrub *Bonellia nervosa*, which has an inverse phenological behavior relative to the majority of tree and shrub species in the dry forest (Chaves and Avalos 2008). In this species, leaf flushing and flower bud production takes place shortly after the autumnal equinox when daylength starts to decrease. The production of a complete canopy in this species happens when the majority of plants are deciduous due to the onset of a very strong dry season. Leaf fall takes place when the forest canopy flushes new leaves before the rains, decreasing light levels in the understory. In this way, *B. nervosa* drops its leaves and remains deciduous when the rest of the plants in the dry forest maintain a full canopy. In the dry forests of Santa Rosa in Costa Rica, photoperiodic variation is only 70 minutes. However, this small variation can induce synchronization of flowering in dry forest species, as well as leaf fall and leaf production (Janzen 1967). Methods for the study of these responses are highly controversial and have not been standardized yet. It has been proposed to study these responses in growth chambers where factors such as light and temperature are highly regulated. However, these responses are expressed at the ecosystem level (Borchert et al. 2005) and are independent of the water conditions of the microenvironment where the plant is located. This shows that factors triggering the response take place at the landscape level, and as such, must include climatic variables like subtle changes in photoperiod. Shifts in the expression of phenological traits will be one of the major expressions of the consequences of climate change (Visser and Both 2005; Cleland et al. 2007).

Variation in the Availability of Water and Its Influence on Plant Primary Production

Tropical latitudes encompass desert environments, seasonal deserts, wetlands, seasonally inundated forests, dry forests, and evergreen, weakly seasonal forests. Due to this diversity of environments, tropical forests show extreme variation in water availability, which has influenced the evolution of diverse physiological strategies to cope with differences in water variation.

One example of a habitat with unique water conditions is the Páramo. These ecosystems exist above the tree line and close to the summit of the highest mountains, showing a combination of physical stress factors, such as excess ultraviolet radiation and very low temperatures, and limited soil organic matter, which combine to produce a highland savanna in which primary production is very limited. Páramos are possibly one of the most unexplored tropical ecosystems in terms of physiological properties, and specifically, regarding biomass production (Rundel et al. 1994). They have different extremes of water availability; this means that these habitats can get almost inundated during the wet season, but experience drought during the dry period. Because of these extremes, the rates of soil organic matter decomposition are slow, especially under waterlogged conditions when organic matter accumulates originating mountain bogs. Páramo and mountain bogs can lose most of their water during dry periods. The available data show that Páramos take a long time to recover lost biomass due to fire and other human disturbances, and thus, represents a very fragile ecosystem (Horn and Kapelle 2009).

African savannas constitute a different type of tropical environment whose primary production is determined by the seasonality and amount of precipitation. Annual precipitation determines 75 percent of the variation in the aboveground primary production of grasses (Oesterheld et al. 1999), mediated through nutrient interactions, especially nitrogen (Hamilton III et al. 1998). Savannas exposed to heavy herbivory from grazers, have better chances to increase primary production on nitrogen-rich soils receiving high precipitation. The influence of fire on primary production is small in comparison to the seasonal movement of large herbivores and their impact on nutrient distribution (McNaughton 1985). The functioning of grassland in the Serengeti is characterized by a high level of energy transfer between trophic levels, which is supported by high alpha diversity in herbivores and plant species, and by fluctuations in resources that are mediated by changes in nutrients and precipitation, which impact the level of herbivore biomass maintained by this dynamic system (Augustine et al. 2003).

Tropical dry forests constitute another tropical environment where water is a critical factor shaping plant production. These environments have been highly disturbed by man. Possibly, the dry tropical forest was the first tropical environment to be almost totally altered by agricultural expansion in the Neotropics (Murphy and Lugo 1986). There are very few samples of pristine tropical dry forests in its mature or primary state. Most of this vegetation is located in regions preferred by humankind for establishing urban centers and agricultural areas; sites that are highly seasonal and where fire could be used as an efficient management tool to control plant regrowth and succession. Thus, most plants within dry forest environments are typical pioneers, and are mostly deciduous species. For this reason, many tropical dry forests are not very productive (Martínez-Yrízar et al. 1996), since they are still in the early stages of regeneration.

The strong seasonality in water availability in tropical dry forests (5–8 months of drought and most of the plant species being deciduous (Jaramillo et al. 2011)) has influenced the development of unique phenological strategies. In addition, plant spatial dispersion is aggregated, which shows that within a species there is selection for specific microenvironments with sufficient access to the water table (Hubbell 1979). Thus, physical factors (in this case water) have a higher importance to community structuring in the dry forests than biological factors (i.e., seed predation and plant competition). Furthermore, in the dry forest there are distinct times for the expression and synchronization of phenological traits. The strong seasonality of water availability, combined with very subtle changes in temperature and photoperiod, influence the synchronization of tree reproductive phenology (see the seminal paper by Janzen 1967). Reproductive phenology is synchronized to a decrease in temperature and photoperiod, which cause morphogenetic responses mediated by phytochromes (Borchert 1994). The end result is the synchronization of reproduction at the landscape level across latitudinal gradients. In dry forest tree species, water can be stored in the bark and in succulent stem tissues. There is a progression of deciduousness depending on the plant capacity to store water. Species that are succulent and with low wood density tend to drop their leaves first, in comparison to species with denser wood and deeper roots. The latter plants can allocate 44 percent or more of their biomass into roots (Clark et al. 2001b). Wood characteristics segregate dry forest tree species in different phenological groups (Brodribb et al. 2002). Dry forest species present more diverse strategies for vegetative and reproductive phenology than any other tropical habitat.

In weakly seasonal environments, such as the rain and moist forest, there are unique adaptations to water availability affecting plant productivity. Specifically, in these habitats plants display root stratification according to susceptibility to water stress (Andrade et al. 2005). Most plants have evolved shallow roots to absorb nutrients from the thin litter layer. However, many life forms like lianas have developed deep roots, which can penetrate up to 30 m (Holbrook and Putz 1996). In the Amazon, the canopy layer can remain evergreen due to the depth of roots, which can reach down to 8 m (Nepstad et al. 1994). Having deep roots facilitates carbon sequestration into the soil. This soil carbon has low turnover rates and remains stored for long periods. It is common for vines and evergreen trees such as figs to develop deep roots, which allow them to access water that would not be available close to the soil surface. Lianas are increasing in abundance across the tropics (Schnitzer and Bongers 2011) and one reason for this increase in dominance is explained by the characteristics of their vascular system. Even though such a system has been considered for many years as atypical, it represents an adaptation to a different strategy of biomass allocation, which emphasizes foliage production in contrast to supporting tissues. Thus, lianas can develop xylem vessels of considerable width, which allows them to have large-scale hydraulic conductivities and draw ground water at high speeds, carrying it over long distances until reaching the canopy surface (Gartner et al. 1990). Lianas have photosynthetic rates similar to many pioneer species, and without investing too much on supporting tissues, they can maintain and replace foliage very rapidly (Avalos et al. 2007). This feature is not trivial, since the lianas target leaf production in the energetically dominant environment of the rainforest canopy. The ability to access deep water sources, the large diameter of xylem vessels, high photosynthetic rates, and better water use efficiently allow lianas to contribute a comparatively larger fraction of the forest photosynthetic biomass (Schnitzer and Bongers 2002). These features make lianas dominant in altered environments, such as forest fragments, forest edges, and secondary forests, as well as the canopy of the primary forest. Lianas represent formidable plants with considerable plasticity, and are destined to become a familiar life form in tropical forests as the process of climate change progresses.

Transfer of Matter and Energy Through an Ecosystem

The transfer of matter and energy takes place through food web interactions. Food web interactions are not necessarily linear and unidirectional as previously suggested by the trophic pyramid concept (Fath and Killian 2007). While the concept of trophic pyramid was useful to consider biomass ratios and for comparing different ecosystems, the relationships among species within an ecosystem overlap, are multidirectional, and involve more than two trophic levels. These interactions rapidly change across space and time. In the tropics, the plants form the base of the food web, but form differentiated modules that are chemically distinct, but functionally very similar. In other words, food webs are separated at the base by plants with different genetic (and chemical) makeup. These separate modules perform similar functions with other trophic levels, such as parasites and herbivores, which tend to be separated closer to the base of the food web, since they have evolved to overcome plant chemical defenses (Gilbert 1980). However, the network structure is functionally similar between food web modules composed by plants of different phylogenetic origin, and hence, different composition of secondary metabolites. Differences in chemical plant quality form the basis of the food web, and keep food networks separated at the base. The interactions between more than two trophic levels connect the upper sections of separated modules (i.e., pollination and seed dispersal). These characteristics are not unique to the food webs in the tropics, but in these environments become extremely complex and represent one of the reasons for the high biodiversity of tropical environments.

Energy and matter flow through the food web superimposed on the functional diversity characteristics of a community. Functional diversity refers to the variety of ecological roles performed by the species of a community. This amplifies the concept of diversity beyond taxonomic composition to include the variety of roles and ecological processes performed by the species (Díaz and Cabido 2001). Diversity, considered as a process and not just as a number, provides a more accurate picture of ecosystem functioning. The direction of the energy flow is also multiple, however, two distinct types of models have been proposed to explain the flow. In the bottom-up model, the energy that flows across the food web at the base (decomposers and producers) determines the complexity of the levels above. This model gives more importance to the chemical differences of the plants forming the base of the food web. In the alternative model, the top-down approach, upper trophic levels control the diversity of the levels below. Specifically, predators control the abundance (and thus, the diversity) of their associated prey, influencing the number of species of the levels below. When top-predators are removed, herbivores are released, affecting the abundance and diversity of target plants. These two models are known as the HSS mechanism after Hairston, Smith, and Slobodkin who published it in their seminal paper of 1960 (Hairston et al. 1960). The HSS hypothesis explains how different trophic levels interact to determine the flow of matter and energy through food webs (Terborgh and Estes 2013). It has been the basis of studies looking at the interaction between species diversity, trophic interactions, and ecosystem stability. Herbivores can impact primary production through selective foraging and the targeting of specific species. Herbivore abundance is determined by the abundance of predators, as well as by competitive interactions among their target plants. The extinction of top predators have direct consequences on plant diversity and plant primary production through their effects on herbivores, and thus, on the competitive balance between target plants. The amount and quality of the plant debris that enter the soil organic matter are also influenced by interactions at the top trophic levels. Top-predators also influence the expression of trophic cascades in terrestrial and aquatic ecosystems (Terborgh and Estes 2013; Pace et al. 1999). For instance, an increase in the density of herbivores decreases net primary production since herbivores consume more plant material, or concentrate on certain species causing the processes of defaunation (or induced species loss). Less organic matter coming from plants affects water and nutrient absorption, thereby affecting the pool of water and nutrients at the ecosystem level. There are also indirect effects on the nitrogen cycle because the quality of plant organic matter entering the forest litter is different relative to the levels previous to the herbivory outbreak. Increased herbivore biomass due to lack of predators and the consequent extinction of the associated plant species in this food web is an example of a trophic cascade. Strong species interactions have profound effects on net primary production through indirect predator-prey effects (Knight et al. 2005). Species interactions determine the abundance, biomass, and production of a community. Trophic cascades are common in aquatic ecosystems while in terrestrial ecosystems they are difficult to detect. In the tropics, there is also a higher degree of trophic specialization relative to temperate areas. This has been proved for Lepidoptera larvae due to the higher diversity of secondary plant compounds in tropical plants, as well as more intense selective pressures from larvae and adult predators. Consequently, tropical caterpillars feed on fewer plant families than their temperate counterparts, in addition to higher diversity of host plant families in the tropics. The end result is a higher number of species diversity at the alpha and beta scales relative to temperate areas (Dyer et al. 2007).

The study of herbivore-plant interactions and their functional consequences at the ecosystem level is an example of the many applications of functional ecology. In this field, taxonomic identities are exchanged with functional roles. While there will never be a consensus regarding how to accurately classify a species in terms of its ecological role, we cannot ignore that many species share a similar function, and that their ecological role has implications for the overall production of a given ecosystem. Trophic networks are characterized by a high degree of redundancy (Lawton and Brown 1994), which influences not only the flow of matter and energy, but also the stability of the food web and its resistance to sudden disturbances. The complexity of trophic interactions demonstrates the importance of understanding the flow of matter and energy in an ecosystem and how it is closely linked with functional diversity.

The practical implications of understanding the relationship between thermodynamics and functional diversity are varied. The take-home message is that ecosystems have evolved over long periods to maintain a dynamic equilibrium with the energy input through the process of photosynthesis. This process is maintained by the sheer number of species that compete and interact in food webs. Altered trophic webs are thermodynamically simple, which will increase the degree of entropy of an ecosystem with the consequent loss of energy. Since biological systems generate much of the energy used by mankind, degraded ecosystems will have a decreased capacity to maintain the diversity of environmental services required to fulfill the basic needs of humankind in the years to come.

Conclusions

Primary production is a complex process driven by abiotic factors (water and nutrient availability, and changes in temperature) and biotic interactions (the capacity of Earth's ecosystems to fix carbon through photosynthesis). The analysis of the limits to primary production has direct links to the wellbeing of humankind.

Human population growth, deforestation, the burning of fossil fuels, the expansion of urban environments, agricultural activity and cattle ranching, and the concentration of human populations in cities, are setting in motion negative feedback mechanisms that are triggering an accelerated global warming. New emerging economies are joining the traditional big economies in a world market dominated by a fast pace of oil-based consumption and hunger for fossil fuel energy. Human impacts on the global climate are clearly disturbing atmospheric composition, with undeniable long-term consequences on the Earth's climate. The increased species extinction, the disturbance and removal of native ecosystems, and the sudden changes in the global composition of the atmosphere and oceans are given rise to a new geologic era termed the Anthropocene.

Tropical ecosystems provide ideal study systems to analyze many of the impacts of climate change on primary production. It is in tropical regions where the interphase atmosphere-biosphere has very intense interactions. It is also in the Tropics where most small economies are concentrated and where the impacts of global warming will be, not only immediate but also highly negative. These economies are not prepared to sustain the impacts of global warming.

Photosynthesis responds in subtle ways to the temperature increases and associated droughts. Species shifts in distribution and composition of the biota are the first signs of adaptation to global warming. The shift is negatively affecting species with slow growth, specialized niches, and low population density. Tropical mountains are among the first environments to show the impacts of global warming. Pollination systems will be the first ones to respond; especially those that are specialized and dependent on immediate energy gain. Tropical latitudes maintain highly diverse and dynamic systems whose response to global warming is still poorly known. More research on the basic responses of these ecosystems to increased temperatures and droughts, as well as shifts in the distribution of biodiversity, will foster better forest management under climate change scenarios, as well as define more specific strategies to improve forest-human interactions from a social and economic perspective. These approaches require the support of the dominant economies, which are the ones responsible for most of the pollution, being the drivers of climate change. The implementation of solutions requires not only sound scientific knowledge, but also ethical responsibility at the global level. These approaches are long overdue.

Acknowledgements

Olivia Sylvester and Carlos Rojas provided comments that improved the original manuscript.

References

ACS (American Chemical Society). 2015. http://www.acs.org/content/acs/en/climatescience/greenhousegases/sourcesandsinks.html

Adger, W.N., S. Huq, K. Brown, D. Conway and M. Hulme. 2003. Adaptation to climate change in the developing world. Prog. Dev. Stud. 3: 179–195.

Andrade, J.L., F.C. Meinzer, G. Goldstein and S.A. Schnitzer. 2005. Water uptake and transport in lianas and co-occurring trees of a seasonally dry tropical forest. Trees 19: 282–289.

Augustine, D.J., S.J. McNaughton and D.A. Frank. 2003. Feedbacks between soil nutrients and large herbivores in a managed savanna ecosystem. Ecol. Appl. 13: 1325–1337.

Avalos, G. and S.S. Mulkey. 1999. Photosynthetic acclimation of the liana *Stigmaphyllon lindenianum* to light changes in a tropical dry forest canopy. Oecologia 120: 475–484.

Avalos, G., S.S. Mulkey, K. Kitajima and S.J. Wright. 2007. Colonization strategies of two liana species in a tropical dry forest canopy. Biotropica. 39: 393–399.

Borchert, R. 1994. Soil and stem water storage determine phenology and distribution of tropical dry forest trees. Ecology 1437–1449.

Borchert, R. and G. Rivera. 2001. Photoperiodic control of seasonal development and dormancy in tropical stem-succulent trees. Tree Physiol. 21: 213–221.

Borchert, R., S.S. Renner, Z. Calle, D. Navarrete, A. Tye, L. Gautier, R. Spichiger and P. von Hildebrand. 2005. Photoperiodic induction of synchronous flowering near the Equator. Nature 43: 627–629.

Brodribb, T.J., N.M. Holbrook and M.V. Gutierrez. 2002. Hydraulic and photosynthetic co-ordination in seasonally dry tropical forest trees. Plant Cell Environ. 25: 1435–1444.

Brokaw, N. and R.T. Busing. 2000. Niche versus chance and tree diversity in forest gaps. Trends Ecol. Evol. 15: 183–188.

Chave, J., C. Andalo, S. Brown, M.A. Cairns, J.Q. Chambers, D. Eamus, H. Fölster, F. Fromard, N. Higuchi, T. Kira, J.P. Lescure, B.W. Nelson, H. Ogawa, H. Puig, B. Riéra and T. Yamakura. 2005. Tree allometry and improved estimation of carbon stocks and balance in tropical forests. Oecologia 145: 87–99.

Chaves, O.M. and G. Avalos. 2008. Do seasonal changes in light availability influence the inverse leafing phenology of the neotropical dry forest understory shrub *Bonellia nervosa* (Theophrastaceae)? Rev. Biol. Trop. 56: 257–268.

Chazdon, R.L. 1988. Sunflecks and their importance to forest understory plants. Adv. Ecol. Res. 18: 1–63.

Chazdon, R.L. and R.W. Pearcy. 1986. Photosynthetic responses to light variation in rainforest species. Oecologia 69: 524–531.

Chazdon, R.L. and R.W. Pearcy. 1991. The importance of sunflecks for forest understory plants. BioScience 41: 760–766.

Clark, D.A., S. Brown, D.W. Kicklighter, J.Q. Chambers, J.R. Thomlinson and J. Ni. 2001a. Measuring net primary production in forests: concepts and field methods. Ecol. Appl. 11: 356–370.

Clark, D.A., S. Brown, D.W. Kicklighter, J.Q. Chambers, J.R. Thomlinson, J. Ni and E.A. Holland. 2001b. Net primary production in tropical forests: an evaluation and synthesis of existing field data. Ecol. Appl. 11: 371–384.

Cleland, E.E., I. Chuine, A. Menzel, H.A. Mooney and M.D. Schwartz. 2007. Shifting plant phenology in response to global change. Trends Ecol. Evol. 22: 357–365.

Díaz, S. and M. Cabido. 2001. Vive la difference: plant functional diversity matters to ecosystem processes. Trends Ecol. Evol. 16: 646–655.

Dietze, M.C. and J.S. Clark. 2008. Changing the gap dynamics paradigm: vegetative regeneration control on forest response to disturbance. Ecol. Monogr. 78: 331–347.

Dyer, L.A., M.S. Singer, J.T. Lill, J.O. Stireman, G.L. Gentry, R.J. Marquis, R.E. Ricklefs, H.F. Greeney, D.L. Wagner, H.C. Morais, I.R. Diniz, T.A. Kursar and P.D. Coley. 2007. Host specificity of Lepidoptera in tropical and temperate forests. Nature 448: 696–699.

Fath, B.D. and M.C. Killian. 2007. The relevance of ecological pyramids in community assemblages. Ecol. Model. 208: 286–294.

Feely, R.A., S.C. Doney and S.R. Cooley. 2009. Ocean acidification: present conditions and future changes in a high-CO_2 world. Oceanography 22: 37–47.

Gartner, B.L., S.H. Bullock, H.A. Mooney, V.B. Brown and J.L. Whitbeck. 1990. Water transport properties of vine and tree stems in a tropical deciduous forest. Am. J. Bot. 77: 742–749.

Gilbert, L.E. 1980. Food web organization and conservation of neotropical diversity. pp. 11–33. *In*: M.E. Soule and B.A. Wilcox (eds.). Conservation Biology: An Evolutionary Ecological Perspective. Sinnauer Asocciates, Sunderland, Massachusetts.

Gower, S.T., C.J. Kucharik and J.M. Norman. 1999. Direct and indirect estimation of leaf area index, f APAR, and net primary production of terrestrial ecosystems. Remote Sens. Environ. 70: 29–51.

Hairston, N.G., F.E. Smith and L.B. Slobodkin. 1960. Community structure, population control, and competition. Am. Nat. 94: 421–425.

Hamilton III, E.W., M.S. Giovannini, S.A. Moses, J.S. Coleman and S.J. McNaughton. 1998. Biomass and mineral element responses of a Serengeti short-grass species to nitrogen supply and defoliation: compensation requires a critical [N]. Oecologia 116: 407–418.

Hansen, J., M. Sato, R. Ruedy, K. Lo, D.W. Lea and M. Medina-Elizade. 2006. Global temperature change. P. Natl. Acad. Sci. USA 103: 14288–14293.

Holbrook, N.M. and F.E. Putz. 1996. Physiology of tropical vines and hemiepiphytes: plants that climb up and plants that climb down. pp. 363–394. *In*: S.S. Mulkey, R.L. Chazdon and A.P. Smith (eds.). Tropical Forest Plant Ecophysiology. Springer US, New York.

Holmgren, M., M. Scheffer, E. Ezcurra, J.R. Gutiérrez and G.M. Mohren. 2001. El Niño effects on the dynamics of terrestrial ecosystems. Trends Ecol. Evol. 16: 89–94.

Horn, S.P. and M. Kappelle. 2009. Fire in the páramo ecosystems of Central and South America. pp. 505–539. *In*: M.A. Cochrane (ed.). Tropical Fire Ecology: Climate Change, Land Use and Ecosystem Dynamics. Springer Berlin, Germany.

Hubbell, S.P. 1979. Tree dispersion, abundance, and diversity in a tropical dry forest. Science 203: 1299–1309.

Janzen, D.H. 1967. Synchronization of sexual reproduction of trees within the dry season in Central America. Evolution 21: 620–637.

Jaramillo, V.J., A. Martínez-Yrízar and R.L. Sanford. 2011. Primary productivity and biogeochemistry of seasonally dry tropical forests. pp. 109–128. *In*: R. Dirzo, H.S. Young, H.A. Mooney and G. Ceballos (eds.). Seasonally Dry Tropical Forests: Ecology and Conservation. Island Press/Center for Resource Economics, Washington D.C.

Keeley, J.E. and P.W. Rundel. 2003. Evolution of CAM and C4 Carbon-Concentrating Mechanisms. Int. J. Plant Sci. 164: S55–S77.

Kitajima, K. and L. Poorter. 2008. Functional Basis for Resource Niche Partitioning by Tropical Trees. pp. 160–181. *In*: W.P. Carson and S.A. Schnitzer (eds.). Tropical Forest Community Ecology. Blackwell, Oxford, United Kingdom.

Knight, T.M., M.W. McCoy, J.M. Chase, K.A. McCoy and R.D. Holt. 2005. Trophic cascades across ecosystems. Nature 437: 880–883.

Lawton, J.H. and V.K. Brown. 1994. Redundancy in ecosystems. pp. 255–270. *In*: E.D. Schulze and H.A. Mooney (eds.). Biodiversity and Ecosystem Function. Springer, Berlin Heidelberg, Germany.

Locatelli, B., V. Evans, A. Wardell, A. Andrade and R. Vignola. 2011. Forests and climate change in Latin America: linking adaptation and mitigation. Forests 2: 431–450.

Losos, E. 1995. Habitat specificity of two palm species: experimental transplantation in Amazonian successional forests. Ecology 76: 2595–2606.

Lüttge, U. 2004. Ecophysiology of crassulacean acid metabolism (CAM). Ann. Bot. London 93: 629–652.

Martínez-Yrízar, A., J.M. Maass, L.A. Pérez-Jiménez and J. Sarukhán. 1996. Net primary productivity of a tropical deciduous forest ecosystem in western Mexico. J. Trop. Ecol. 12: 169–175.

McNaughton, S.J. 1985. Ecology of a grazing ecosystem: the Serengeti. Ecol. Monogr. 55: 259–294.

Murphy, P.G. and A.E. Lugo. 1986. Ecology of tropical dry forest. Annu. Rev. Ecol. Syst. 17: 67–88.

Nepstad, D.C., C.R. de Carvalho, E.A. Davidson, P.H. Jipp, P.A. Lefebvre, G.H. Negreiros and S. Vieira. 1994. The role of deep roots in the hydrological and carbon cycles of Amazonian forests and pastures. Nature 372: 666–669.

New, M., D. Liverman, H. Schroder and K. Anderson. 2011. Four degrees and beyond: the potential for a global temperature increase of four degrees and its implications. Philos. T. R. Soc. A 369: 6–19.

Nielsen, S.L., S. Enriquez, C.M. Duarte and K. Sand-Jensen. 1996. Scaling maximum growth rates across photosynthetic organisms. Funct. Ecol. 10: 167–175.

NOAA. 2015. http://www.esrl.noaa.gov/gmd/ccgg/trends/#mlo_data

Oesterheld, M., J. Loreti, M. Semmartin and J.M. Paruelo. 1999. Grazing, fire, and climate effects on primary productivity of grasslands and savannas. pp. 287–305. *In*: L.L. Walker (ed.). Ecosystems of Disturbed Ground. Elsevier, Amsterdam.

Otárola, M.F. and G. Avalos. 2014. Demographic variation across successional stages and their effects on the population dynamics of the neotropical palm *Euterpe precatoria*. Am. J. Bot. 101: 1023–1028.

Pan, Y., R.A. Birdsey, J. Fang, R. Houghton, P.E. Kauppi, W.A. Kurz, O.L. Phillips, A. Schvidenko, S.L. Lewis, J.G. Canadell, P. Ciais, R.B. Jackson, S.W. Pacala, A.D. McGuire, S. Piao, A. Rautiainen, S. Sitch and D. Hayes. 2011. A large and persistent carbon sink in the world's forests. Science 333: 988–993.

Riebesell, U., K.G. Schulz, R.G.J. Bellerby, M. Botros, P. Fritsche, M. Meyerhöfer and E. Zöllner. 2007. Enhanced biological carbon consumption in a high CO_2 ocean. Nature 450: 545–548.

Rivera, G. and R. Borchert. 2001. Induction of flowering in tropical trees by a 30-min reduction in photoperiod: evidence from field observations and herbarium specimens. Tree Physiol. 21: 201–212.

Rundel, P.W., A.P. Smith and F.C. Meinzer. 1994. Tropical Alpine Environments: Plant Form and Function. Cambridge University Press, United Kingdom.

Sage, R.F. 2004. The evolution of C4 photosynthesis. New Phytol. 161: 341–370.

Schnitzer, S.A. and F. Bongers. 2002. The ecology of lianas and their role in forests. Trends Ecol. Evol. 17: 223–230.

Schnitzer, S.A. and F. Bongers. 2011. Increasing liana abundance and biomass in tropical forests: emerging patterns and putative mechanisms. Ecol. Lett. 14: 397–406.

Silvertown, J. 2004. Plant coexistence and the niche. Trends Ecol. Evol. 19: 605–611.

Sine, K.A.A. 2002. The hot and the classic. Plant Physiol. 10: 900020.

Smith, B.D. and M.A. Zeder. 2013. The onset of the Anthropocene. Anthropocene 4: 8–13.

Terborgh, J. and J.A. Estes. 2013. Trophic Cascades: Predators, Prey, and the Changing Dynamics of Nature. Island Press, Washington D.C.

Visser, M.E. and C. Both. 2005. Shifts in phenology due to global climate change: the need for a yardstick. P. R. SOC. B 272: 2561–2569.

Walker, L.R., D.A. Wardle, R.D. Bardgett and B.D. Clarkson. 2010. The use of chronosequences in studies of ecological succession and soil development. J. Ecol. 98: 725–736.

Wright, S.J. 1996. Phenological responses to seasonality in tropical forest plants. pp. 440–460. *In*: S.S. Mulkey, R.L. Chazdon and A.P. Smith (eds.). Tropical Forest Plant Ecophysiology. Springer-Verlag, New York.

Wright, S.J. and O. Calderón. 2006. Seasonal, El Niño and longer term changes in flower and seed production in a moist tropical forest. Ecol. Lett. 9: 35–44.

CHAPTER 3

Forest Carbon Sequestration and Global Change

Achim Häger[1,]* and *Luitgard Schwendenmann*[2]

ABSTRACT

Forests contain 70–80 percent of terrestrial carbon (C) and represent a global C sink, despite of continued deforestation. On the other hand, forests and land-use change (LUC) account for the largest uncertainties within the global C balance. The future role of forests is even more uncertain within the context of global change. The question if forests will shift from C sinks to sources is fundamental for conservation and management. We aim to synthesize current knowledge on the role of the major forest biomes in the global C cycle. Furthermore, we evaluate management options, based on their potential to maximize greenhouse gas (GHG) mitigation and other benefits from forest ecosystems. Whereas boreal and temperate forests currently act as net sinks, there is a potential for increased insect outbreaks, fires, and droughts. Pressure from infrastructure development and the demand for biomass fuels is increasing. Tropical forests account for the largest gross C sink, but face deforestation and degradation. The future role of tropical forests depends mostly on LUC dynamics. Uncertainties persist about deforestation rates and impacts from global change. It

[1] School for Field Studies, Center for Sustainable Development Studies, P.O. Box 150-4013, Atenas, Alajuela, Costa Rica.
E-mail: ahaeger@fieldstudies.org
[2] School of Environment, The University of Auckland, Private Bag 92019, Auckland 1142, New Zealand.
E-mail: l.schwendenmann@auckland.ac.nz
* Corresponding author

is clear, however, that changes in the C balance of tropical forests will affect atmospheric CO_2 levels on a global scale. In managed temperate and boreal forests, aboveground carbon (AGC) can be maximized by optimizing rotation cycles in even-aged stands. Mixed-species stands may increase productivity and resilience to disturbances. Research is needed on the impacts of silvicultural activities on soil organic carbon (SOC). In the tropics, reforestation and deforestation avoidance are the most important GHG mitigation pathways. Lessons learnt from REDD (reduced emissions from deforestation and forest degradation) pilot projects may help to establish GHG mitigation strategies in the future. The majority of timber harvest in the tropics is unsustainable and could be improved by extending rotation cycles, reduced impact logging, and long-term management planning. International support is needed to ensure capacity building and strong institutions, and to create incentives for global markets. The extension of planted forests is growing worldwide. Plantations sequester important amounts of C, although afforestation may initially lead to SOC losses, and AGC storage over multiple rotation cycles is low, compared to natural forests. Conservation of old-growth forests is the most effective way to secure C storage.

Introduction

Forests cover approximately 40 million km^2, or 31 percent of the global land surface area (FAO 2011).[1] Almost a third of the original forest cover has been lost due to human activities (FAO 2012). Whereas forests in boreal and temperate biomes are currently expanding, tropical forest area is decreasing. The global annual net deforestation rate between 2000 and 2010 was 0.14 percent or 52,110 km^2 (roughly the size of Costa Rica). Thirty-six percent of the current forested area is relatively undisturbed (primary forest), 57 percent is naturally regenerated secondary forest, whereas the remaining 7 percent is planted (FAO 2010; FAO 2011).

Forests play a crucial role in the global carbon (C) cycle, as they contain 70–80 percent of terrestrial C (Baccini et al. 2012). Global C storage in forest biomass and soils to 1 m depth ranges between 800 and 1400 Pg. Despite net deforestation rates, forests continue to be a C sink, removing 14 percent of annual anthropogenic C emissions from fossil fuels and cement production (Robinson 2007; Pan et al. 2011; Reich 2011; Houghton et al. 2012). However, there are still considerable uncertainties regarding estimates of forest cover change, C density in different land-use types, or C storage in soils

---

[1] The FAO (2010) defines forests as land areas > 0.5 ha with trees higher than 5 m and a canopy cover > 10 percent. The definition does not include land under agricultural or urban use.

(e.g., Lorenz and Lal 2010; Houghton et al. 2012; Kim et al. 2015). Contradictory information exists about the total C stocks and the allocation of C in different pools across forest biomes, which can partly be explained by ambiguous definitions for different biomes (e.g., Watson et al. 2000; Sabine et al. 2004; Pan et al. 2011; Tyrrell et al. 2012). The future role of forests in the global C cycle is even more uncertain, due to complex and non-linear feedback mechanisms among altered biogeochemical cycles, changing climate and the biosphere. Impacts from drought, extreme meteorological events, rapid temperature increase, fire or insect outbreaks could increase C release across forest biomes (e.g., Heimann and Reichstein 2008; IPCC 2014). The question as to whether forests are likely to shift from C sinks to sources is fundamental for conservation and management, as well as for greenhouse gas (GHG) mitigation strategies.

The objective of this chapter is to synthesize our current knowledge on the role of forests in the global C cycle and to analyze GHG mitigation potentials in the context of global change and complex feedback mechanisms for the major forest biomes. Furthermore, management options will be evaluated based on their potential to maximize C storage and other benefits from forest ecosystems.

The Role of Forest Biomes in the Global Carbon Cycle

Forest Carbon Fluxes and Stocks on a Global Scale

As C is the main element of organic matter, the biosphere plays a major role in regulating the global C cycle (Chapin et al. 2006; Heimann and Reichstein 2008). Forests account for more than 90 percent of global terrestrial biomass, containing over 80 percent of terrestrial aboveground carbon (AGC) and over 70 percent of belowground C (Jandl et al. 2007; Lorenz and Lal 2010; Pan et al. 2013). Recent estimates for total C storage in forest vegetation and soils to 1 m depth vary between 861 Pg (Pan et al. 2011) and 1659 Pg (Lorenz and Lal 2010). Pan et al. (2011) reported that 44 percent of forest C is stored in soils, 42 percent in living biomass, and 13 percent in dead organic matter.

Photosynthesis represents the most important C flux into terrestrial ecosystems. Estimated global annual uptake by forests (= gross primary production, GPP) ranges between 59 and 75 Pg C/yr, which represents about 50–60 percent of total global GPP by vegetation (Denman et al. 2007; Beer et al. 2010; Lorenz and Lal 2010; Pan et al. 2013). The growth and maintenance of plant tissue requires energy, which is lost to autotrophic respiration

(R_a) (Fig. 1). The remaining amount is defined as net primary production (NPP = GPP – R_a). Global forest NPP lies around 30 Pg C/yr (Saugier et al. 2001; Sabine et al. 2004):

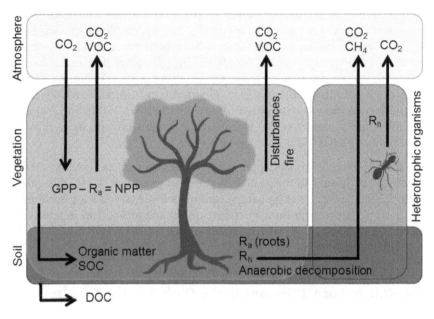

Fig. 1. Simplified forest carbon cycle (GPP = gross primary production, NPP = net primary production, R_a = autotrophic respiration, R_h = heterotrophic respiration, SOC = soil organic carbon, DOC = dissolved organic carbon, VOC = volatile organic compounds).

(Source: information from Prentice et al. 2001; Luyssaert et al. 2007; Lorenz and Lal 2010).

NPP is allocated to biomass within different plant structures. The partitioning of NPP is relevant to C cycling, because residence times of C differs, e.g., from months in foliage and fine roots to centuries in stems (Luyssaert et al. 2007).

Carbon stored in biomass and soils is returned to the atmosphere by R_a, heterotrophic respiration (R_h) from decomposers and herbivores, and by fire (Fig. 1). Dead organic matter is transferred into litter and soil layers. Residence time of C can range from months in litter to millennia in soil organic carbon (SOC) (Prentice et al. 2001; Luyssaert et al. 2007; Lorenz and Lal 2010). The balance between litter accumulation and decomposition is mostly controlled by the quality of organic matter and by climate. The stabilization of SOC depends mainly on soil properties, such as the presence of clay minerals that form stable compounds with organic matter (Jandl et al. 2007).

The remaining fraction of NPP is defined as annual net ecosystem production (NEP):

$$NEP = NPP - R_h$$

NEP can be estimated by measuring C stock changes in vegetation and soil. The amount of terrestrial NEP has been projected to be close to 10 Pg/yr (Prentice et al. 2001). However, GPP and total ecosystem respiration ($R_e = R_a + R_h$) alone do not represent a balance. The actual net rate of C accumulation or loss within an ecosystem over time (dC/dt) is determined by the net ecosystem C balance (NECB), which includes additional C fluxes between ecosystems and the atmosphere, such as non-CO_2 C fluxes and lateral C transfer between ecosystems, e.g., by wind or water (Chapin et al. 2006; Luyssaert et al. 2007; Chapin et al. 2009; Lorenz and Lal 2010):

$$NECB = -NEE + F_{CO} + F_{CH4} + F_{BVOC} + F_{DIC} + F_{DOC} + F_{PC}$$

$-NEE$, the net ecosystem exchange, describes the net CO_2 flux from an ecosystem to the atmosphere (NEE < 0 equals CO_2 uptake), and can be measured by eddy covariance techniques (e.g., Chapin et al. 2006). Over short time scales NEE, NEP, and NECB converge closely; however, over large temporal and spatial scales, NECB is more strongly influenced by inorganic C and non CO_2 C fluxes (Chapin et al. 2006; Luyssaert et al. 2007; Chapin et al. 2009).

F_{CO} describes the net carbon monoxide (CO) exchange by plants and ecosystems, a flux that remains to be understood in detail (Denman et al. 2007; Lorenz and Lal 2010).

F_{CH4} is the net exchange of methane (CH_4). CH_4 is principally released by anaerobic decomposition processes (Fig. 1) and to a lesser extent by living vegetation. Meanwhile, CH_4 oxidation by soil bacteria represents an important CH_4 sink (Keppler et al. 2006; Denman et al. 2007; Lorenz and Lal 2010). Forests account for about 50 percent of the global CH_4 soil sink, with an uptake of 11.6 Tg/yr (Dutaur and Verchot 2007).

F_{BVOC} is the flux of biogenic organic volatile compounds (BVOC). Globally, isoprene is the most important VOC emitted by vegetation. Isoprene is released at a rate of 0.5–0.8 Pg/yr, which equals 0.4–0.7 Pg C/yr, a value that is significant in comparison to global NEP (Lorenz and Lal 2010). Additionally VOCs affect atmospheric chemistry, e.g., the formation of tropospheric ozone (Guenther et al. 2006; Denman et al. 2007).

F_{DOC} and F_{DIC} describe fluxes of dissolved organic and inorganic carbon. These processes include transfer of dissolved C between ecosystems or C losses from leaching (Chapin et al. 2006).

F_{PC} refers to fluxes of particulate C (PC), resulting, e.g., from erosion processes, wind or soot from fires (Chapin et al. 2006).

Impact of Land-Use Change on the Global Carbon Cycle

Land-use change (LUC) was the dominant source of C emissions until the early 20th century and still accounts for more than 25 percent of historical accumulative emissions (Le Quéré et al. 2014; Global Carbon Project 2014). Currently, LUC represents the second largest CO_2 source after fossil fuel consumption (Le Quéré et al. 2009; Friedlingstein et al. 2010; Houghton et al. 2012). Between 2000 and 2009 LUC was responsible for 12.5 percent of anthropogenic CO_2 emissions, equivalent to 1.1 Pg C/yr (Friedlingstein et al. 2010; Pan et al. 2011; Houghton et al. 2012). The importance of LUC emissions is declining, both in relative and absolute terms, due to increasing fossil fuel emissions and decreasing deforestation rates. In 2013, CO_2 emissions from LUC accounted only for 8 percent of total global emissions (Global Carbon Project 2014). Overall, uncertainties about LUC C fluxes are greater than those associated with fossil fuel emissions (Le Quéré et al. 2009; Friedlingstein et al. 2010).

The gross C sink in forests amounts to 4.1 ± 0.7 Pg/yr. The largest proportion of C is absorbed by living biomass (72 percent) and forest soils (12 percent) (Pan et al. 2011). Despite net deforestation rates, the world's forests represent a net sink. Forest biomes constitute 50–60 percent of the total terrestrial C sink and remove between 1.1 and 2.4 Pg C/yr from the atmosphere, due to processes such as forest regrowth, responses to increased atmospheric CO_2 levels, extended growth seasons in temperate and boreal areas, as well as nitrogen (N) depositions (Grace 2004; Robinson 2007; Lorenz and Lal 2010; Pan et al. 2011; Reich 2011; Le Quéré et al. 2014).

Boreal Forests

Boreal forests cover an area of approximately 15 million km², or 15 percent, of the earth's terrestrial surface (Gower et al. 2001) (Table 1). They represent the second largest forest biome and are located between 50° and 70° N. The climate is characterized by short, mild summers (+30°C temperature maxima) and long winters (–60°C temperature minima) (Larsen 1980). Low air and soil temperatures exert a strong influence on tree species composition, productivity, and decomposition. Annual precipitation in the boreal region ranges between 300 and more than 1,000 mm, depending on the distance to the nearest ocean (Müller 1982).

Table 1. Estimates of C stocks and sinks in the boreal forest biome.

Source	Area (million km²)	Biomass C Pg	SOC 0–1 m Pg	net C sink Pg/yr
Pan et al. (2011)	11.4	54	175	0.50
Lorenz and Lal (2010)	9.5–14.7	78–143	338	0.49–0.70
Sabine et al. (2004)	13.7	57	112	-
Prentice et al. (2001)	13.7	88*	471	-

*Value is likely to an overestimation due to estimates from Russia that include standing dead biomass.

Boreal forests are often underlain by discontinuous permafrost (permanently frozen ground). Soils are a mosaic of sandy mineral soils and organic peat (DeLuca and Boisvenue 2012). Over 30 percent of the soils in the boreal biome are Spodosols, with a high content of organic matter, and Gelisols as defined by USDA Soil Taxonomy (Soil Survey Staff 1999).

Dominant tree species include spruce (*Picea spp.*), pine (*Pinus spp.*), fir (*Picea spp.*) and, in the case of Eurasian forests, also larch (e.g., *Larix sibirica* Ledeb. and *Larix gmelinii* (Rupr.) Kuzen.). In the understory ericaceous shrubs and bryophytes such as feather moss (*Pleurozium spp.*) and sphagnum (*Sphagnum spp.*) predominate, depending on soil moisture conditions (Gower et al. 2003). Wildfire is the dominant natural driver of species composition, floristic diversity, and forest dynamics (Soja et al. 2007). Early successional stages after fire are dominated by the deciduous tree genera *Alnus*, *Betula*, and *Populus* (Black et al. 2005).

Boreal forests play an important role in the global C cycle as they store between 18 and 30 percent of total terrestrial C (Watson et al. 2000; Robinson 2007; DeLuca and Boisvenue 2012). Biomass C stocks in boreal forests range between 54 and 143 Pg (Table 1). SOC estimates range between 112 and 471 Pg in the top 1 m of soil (Table 1). Boreal peatlands account for an additional 260 to 600 Pg C (Gorham 1991; Kasischke and Stocks 2000). The low temperatures lead to the accumulation of deep organic soil layers. SOC exceeds the C pool in the vegetation (De Deyn et al. 2008).

The amount of C stored in aboveground biomass (AGB) varies widely, depending on stand age, forest structure, and region. The AGC stocks in Canadian boreal forests at various stages of succession (1–154 years in age) ranged from 50 to 110 Mg/ha (Goulden et al. 2011). In mature (150–300 years) stands of *Abies*, *Pinus*, *Picea,* and *Larix* in Siberia, total AGC varied between 50 and 130 Mg/ha, depending on the time since the last fire (Schulze et al. 2012). Luyssaert et al. (2007) reported an AGC between 48 Mg/ha for boreal semiarid forests and 76 Mg/ha for humid forest types. Pan et al. (2013) distinguished between coniferous boreal forests, mountain forests, and tundra woodlands and reported living biomass (above- and belowground) C densities of 48, 19, and 7 Mg/ha for these vegetation types, respectively. Belowground C ranges between 25 and 55 Mg/ha, although the root C pool is most likely underestimated (IPPC 2007; Robinson 2007; Keith et al. 2009).

The contribution of mosses and bryophytes is specific to the boreal region and can be up to 20 percent of AGC in older forest stands (Goulden et al. 1998). Bryophyte communities have important effects on SOC storage, as they decompose slowly and thus tend to accumulate between fire events (Turetsky et al. 2010). Furthermore, the moss layer reduces heat penetration, which promotes the development of permafrost (Startsev et al. 2007).

Between 130 to 450 Mg C/ha are stored in the top 1 m of North American boreal forests (Ping et al. 2010). In Alaskan boreal forests C stocks (0–1 m) varied significantly in the mineral soil (70–290 Mg/ha) and organic layers (13–86 Mg/ha), depending on topography, aspect, tree species, stand age, and permafrost presence (Johnson et al. 2010). Charcoal, which is considered a comparatively stable form of C, accounts for 1–50 percent of total C in boreal forest soils (Schulze et al. 1999; DeLuca and Aplet 2008; Guggenberger et al. 2008). The quantity and quality of charcoal depends on numerous factors including the severity and frequency of fire and opportunities for charcoal to mix into mineral soil layers (Certini 2005).

With 8.3 Pg/yr, GPP is comparably low in the boreal forest biome, due to low temperatures, a short growing season, and low soil mineral N supplies (Litton et al. 2007; Beer et al. 2010). Average GPP is estimated at 8–12 Mg C/ha/yr (Luyssaert et al. 2007). NPP accounts only for 25–30 percent of GPP (Trumbore 2006). Aboveground and belowground R_a losses range between 1.3 and 3.8 Mg C/ha/yr for boreal tree species (Litton et al. 2007). Based on eddy covariance flux estimates and direct measurements, NPP for boreal forest types ranges between 2 and 15 Mg C/ha/yr (Pregitzer and Euskirchen 2004; Luyssaert et al. 2007; Nieder and Benbi 2008). NPP might be underestimated, as inventories and measurements may not completely account for understory and bryophyte vegetation, fine root production, root exudates, mycorrhiza, reproductive organs, herbivory, tree mortality, and emissions of VOC and CH_4 (Gower et al. 2001; Luyssaert et al. 2007). The productivity of bryophytes in boreal forests can be equal to or exceed the stem growth of trees (Gower et al. 2001). Czimczik et al. (2006) found that bryophytes made up 20 percent of total aboveground NPP in a Canadian black spruce stand. Several studies have found that C production efficiency (NPP/GPP) declines in older boreal stands due to increased R_a (Goulden et al. 2011).

Estimates of losses by R_h and R_e among boreal forests range between 1.5–3.8 Mg C/ha/yr and between 7.3–10.3 Mg C/ha/yr, respectively (Pregitzer and Euskirchen 2004; Luyssaert et al. 2007). Pregitzer and Euskirchen (2004) estimated a mean NEP of 0.3 Mg C/ha/yr for boreal forests. Higher NEP estimates (0.4–1.8 Mg C/ha/yr) have been reported by Luyssaert et al. (2007).

The total C sink of the boreal forest biome (in plants and soil) is estimated to range between 0.49 and 0.70 Pg/yr (Apps et al. 1993; Robinson 2007; Lorenz and Lal 2010; Pan et al. 2011) (Table 1). This is about one-third of the total northern land C uptake, estimated at 1.5 Pg/yr (Stephens et al. 2007).

Overall, boreal forest area has remained fairly stable between 1990 and 2010 (MCPFE 2007; FAO 2010). Fire and insect outbreaks are the dominant natural disturbances influencing NPP and C balance (Peng and Apps 1999; Wirth et al. 2002; Stocks 2004; Bond-Lamberty et al. 2007; Bergeron and Harper 2009). Overall, an estimated 5–15 million ha burn annually in the boreal biome (Kasischke and Stocks 2000). Natural fire cycles ranges from 50 to 200 years, but fire return intervals have increased because of human interference (Stocks et al. 1996). Carbon loss by fire ranges between 35 and 85 Tg/yr for boreal North America and between 58 and 273 Tg/yr for boreal Russia and Siberia (Balshi et al. 2007). Fires release CO_2 as well as smaller amounts of CO and CH_4. Between 1 and 7 percent of burning biomass is converted into charcoal (Preston and Schmidt 2006). However, it is debatable whether charcoal is a potential C sink in boreal forests by contributing to the stable SOC fraction (Czimczik and Masiello 2007; Zimmermann et al. 2012).

The C balance of boreal forests is also affected by insect-related disturbances (Bernier and Apps 2006). In particular, photosynthetic C uptake is reduced through defoliation and R_e is enhanced by insect respiration and accelerated decomposition (Kurz et al. 2008; Edburg et al. 2012).

Temperate Forests

Temperate forests cover between 7.7 and 14.2 million km^2, representing one quarter to one third of the global forest area (Table 2). Estimates vary, because definitions along the borders of boreal and subtropical biomes are ambiguous (see e.g., Luyssaert et al. 2007; Tyrrell et al. 2012; Pan et al. 2013). Most temperate forests (including subtropical areas) are located between 25° N and 55° N. The Southern hemisphere has only small areas of temperate forests, e.g., in South America, Australia, and New Zealand. Temperatures may vary between –30°C and 30°C and in general are strongly seasonal. Throughout the biome, precipitation ranges between 500 mm to over

Table 2. Ranges for estimates of C stocks and sinks in the temperate forest biome.

Source	Area (million km^2)	Biomass C Pg	SOC 0–1 m Pg	net C sink Pg/yr
Pan et al. (2011)	7.7	47	57	0.78
Lorenz and Lal (2010)	10.4–14.2	73–159	153–195	0.37
Sabine et al. (2004)	10.4	139	195	-
Prentice et al. (2001)	10.4	59–139	100–153	-

3500 mm (Lorenz and Lal 2010; Pan et al. 2011; Tyrrell et al. 2012; Pan et al. 2013). Almost 90 percent of the forests occur under relatively humid conditions, while the rest are found in Mediterranean climate types with more seasonal rainfall patterns (Lorenz and Lal 2010; Tyrrell et al. 2012).

After a long history of land-use change, resource exploitation, and urban development, temperate forests cover only about half of their original extent and may have lost about 70 percent of their original biomass. They are generally intensively managed, logged over, fragmented, or planted. Only 1–3 percent of the remaining temperate forests are considered to be relatively undisturbed (FAO 2010; Lorenz and Lal 2010; Tyrrell et al. 2012; Pan et al. 2013).

Temperate forests store the least C across all forested biomes, although estimates overlap with boreal forests (Tables 1–3) (Prentice et al. 2001; Sabine et al. 2004; Lorenz and Lal 2010; Pan et al. 2011). The highest amounts of living biomass (208 Mg/ha) are stored in oceanic temperate forests with high rainfall and a continuous growing season, whereas the lowest amounts are found in mountain systems (59 Mg/ha) and continental temperate forests (61 Mg/ha). Subtropical forests have intermediate biomass C densities (66–77 Mg/ha) (Pan et al. 2013). Estimated total C stocks range between 104 and 354 Pg, which is equivalent to 15–20 percent of C stored in forests globally (Table 2). Tyrrell et al. (2012) reported that temperate forests accumulate 11 percent of the global terrestrial C stocks, based on a lower range of estimates (99–159 Pg), compiled from different sources.

SOC is the most important C pool, comprising between half and two thirds of total C, while the rest is stored in biomass (Table 2). Tyrrell et al. (2012) argued that SOC may be greatly underestimated, because knowledge about SOC at depths below 1 m remains limited. In general, deciduous forests store higher amounts of SOC in 0–3 m depth (174 Mg/ha), than temperate evergreen forests (146 Mg/ha) (Jobbagy and Jackson 2000). Around 26 percent of total SOC in temperate forests is stored between 1 and 3 m in depth (Jobbagy and Jackson 2000). Pan et al. (2011) estimated that 39 percent of C is stored in living biomass, 13 percent in leaf litter and coarse woody debris, and 48 percent in SOC to a depth of 1 m. Belowground biomass may account for 5–10 percent of total C; however, fine root biomass is difficult to measure and likely to be underestimated (Robinson 2007; Lorenz and Lal 2010; Tyrrell et al. 2012).

C storage varies considerably on a regional scale across different forest types and successional stages. Tyrrell et al. (2012) compiled estimates from 18 different sites, ranging between 58 and 1013 Mg C/ha stored in living biomass and soils at varying depths (0.2–1 m). The highest value came from an old-growth coniferous rainforest in Oregon. The highest known C density in the world was measured in Australian old-growth *Eucalyptus regnans* F. Muell., with an average of 1867 Mg/ha, taking into account only aboveground living and dead biomass (Keith et al. 2009). Other temperate tree species with extraordinary high C storage capacity include *Sequioa sempervirens* Endl. and *Pseudotsuga menziesii* (Mirb.) Franco (Lorenz and Lal 2010). Warm temperate moist (subtropical) and cool temperate moist forests store very high amounts of biomass C (on average 498 and 642 Mg/ha, respectively), due to climatic conditions that allow for high productivity and slow decomposition rates, in combination with minimal human disturbance (Keith et al. 2009).

GPP of the temperate biome is 9.9 Pg C/yr, or roughly 8 percent of annual global GPP. This amount lies just above the average for boreal forests and represents about a quarter of the GPP estimated for tropical forests (Beer et al. 2010). Temperate humid evergreen forests have the highest GPP in this biome (17.6 Mg C/ha/yr, roughly half the value estimated for tropical humid forests), followed by Mediterranean evergreen (14.8 Mg C/ha/yr) and temperate deciduous forests (13.8 Mg C/ha/yr). The temperate semi-arid forest is the least productive with 12.3 Mg C/ha/yr (Luyssaert et al. 2007).

Total biome NPP estimates range between 7.4 and 8.1 Pg C/yr (Sabine et al. 2004; Beer et al. 2012). Average NPP per unit area lies at 6.7 Mg C/ha/yr. NPP in Mediterranean evergreen forests is almost as high as in tropical humid forests with 8.0 Mg C/ha/yr, whereas the semi-arid evergreen forest type shows the lowest productivity (3.5 Mg C/ha/yr) (Luyssaert et al. 2007). Across the temperate biome, a major part of NPP is allocated to wood production (36–49 percent), except for the semi-arid evergreen forests, where nearly half of NPP is invested into root biomass (Luyssaert et al. 2007). Average R_a is lower than in tropical forests, ranging from 4.0–9.5 Mg C/ha/yr (41–54 percent of GPP, respectively) in different temperate forest types. Consequently, the NPP/GPP ratio (production efficiency) in temperate forests can be more than twice as high as in tropical forests (54 percent for Mediterranean evergreen and humid deciduous, compared to 24 percent in tropical humid forests) (Luyssaert et al. 2007). Based on MODIS remote sensing data, Zhang et al. (2009) found that NPP/GPP is lower in wet and warm climates. They further showed that NPP/GPP is lower in broadleaf evergreen forest compared to deciduous forests, and lower in broadleaf compared to needle forest types.

Carbon losses from R_e range between 11.0 Mg C/ha/yr (semi-arid evergreen) and 13.4 Mg C/ha/yr (humid evergreen), or less than half of the amount estimated for tropical forests (Luyssaert et al. 2007). Consequently, the NEP of 4.0 Mg C/ha/yr for humid evergreen forests is almost identical to tropical humid forests. The lowest NEP was found in the semi-arid forest type (1.3 Mg C/ha/yr). Tyrrell et al. (2012) compiled data on NEP from 15 different studies of European and North American forests. These stands ranged from being C sources (–1.2 Mg/ha/yr for a 20-year old *Pinus ponderosa* P. Lawson & C. Lawson stand) to strong C sinks (7.2 Mg/ha/yr for a young *Pinus taeda* L. plantation). The strong NEP variations can partly be explained by management interventions in temperate forests (Lorenz and Lal 2010; Tyrrell et al. 2012).

On a global scale, temperate forests currently represent a sink of 0.2–0.8 Pg C/yr, comprising approximately a third of the global forest sink (Pan et al. 2011; Tyrrell et al. 2012) (Table 2). Furthermore, temperate forests on well-drained soils take up 3.3 Tg CH_4/yr, which is the highest rate among forest biomes (Dutaur and Verchot 2007).

Temperate forest cover increased by 5 percent between 1990 and 2007 (Pan et al. 2011). During this period the biomass sink augmented by 17 percent, due to increasing forest area and C density per area, for example, in China and North America. The recent large-scale recovery of forest area results from the abandonment of agricultural land and afforestation activities. Large-scale deforestation is not expected in the near future. Currently a large proportion of middle aged (50–100 years) temperate forest stands are contributing to sustained high C sequestration rates, but this sink may eventually decline when these forests mature (Bonan 2008; Fahey et al. 2010; Pan et al. 2011; Tyrrell et al. 2012).

The status of temperate forests as a small C sink is vulnerable to the accumulative impact from pollutants (e.g., NO_x), species shifts, or a climate change induced increase in droughts or disturbances. Only small changes in forest cover or age class distributions could shift temperate forests from C sinks into sources (Pan et al. 2011; Tyrrell et al. 2012). The C sink of European forests may already be declining as a result of slowing stem increment in maturing forest stands, decreasing reforestation rates due to infrastructure development, and increasing disturbances (Nabuurs et al. 2013). Environmental change in combination with increasing demands from a growing human population may put new pressures on temperate forest resources after the recent period of recovery (Bonan 2008). For instance,

Buongiorno et al. (2011) projected a decrease of up to 2–9 percent in standing forest stocks for countries in the European Union, the United States, and China, due to an increased demand for bioenergy.

Tropical Forests

According to different estimates and definitions, tropical forests cover between 14.5 and 22.0 million km² (10–15 percent) of the land surface area (Table 3). They represent the largest forest biome, comprising almost 50 percent of the global forest cover. The largest extensions exist in the Amazon and Congo basins (Lorenz and Lal 2010; Mahli 2010; Pan et al. 2011). Tropical forests are located between 23.5° N and 23.5° S; temperatures vary between 20–25°C and annual rainfall exceeds 2000 mm, except for dry forests (Murphy and Lugo 1986; Lorenz and Lal 2010). Climate, soil conditions, topography, and species composition vary greatly, which leads to significant differences in C fluxes across this biome. The FAO (1993) classified tropical forests into rainforests (41 percent of forest cover in the tropics), moist deciduous forests (33 percent), dry forests (14 percent), and mountain forests (12 percent). Current land cover area differs greatly from the original extent of these forest types. For instance, tropical dry forest may have covered more than 40 percent of the tropics and subtropics, but most of it has been converted to agriculture or degraded (Murphy and Lugo 1986). The remaining tropical forests vary in terms of their condition and management intensity. Fifty-three percent of the forested area in the tropics is classified as primary forest without significant disturbance (Blaser et al. 2011). General trends show decreasing primary forests and increasing areas of plantations and forests affected by human activity (FAO 2010). On the other hand about 40 percent of the world's remaining relatively undisturbed forests are found in the tropics (Lorenz and Lal 2010). Better estimates on the distributions of primary, secondary, and disturbed forests are needed to assess and predict C fluxes in the tropics accurately (Meister et al. 2012). Furthermore, there are few studies on C fluxes in tropical dry and montane forests, although they differ considerably from lowland tropical humid forests (Mahli 2010; Meister et al. 2012).

Table 3. Ranges for estimates of C stocks and sinks in the tropical forest biome.

Source	Area (million km²)	Biomass C Pg	SOC 0–1 m Pg	net C sink Pg/yr
Pan et al. (2011)	19.5	262	151	–1.1
Lorenz and Lal (2010)	14.5–22.0	206–389	214–435	0.72–1.30
Sabine et al. (2004)	17.5	340	435	-
Prentice et al. (2001)	17.5–17.6	212–340	213–216	–1.1–1.3

Tropical forests account for 20–40 percent of the global terrestrial C pool (Prentice et al. 2001; Sabine et al. 2004; Grace et al. 2006; Meister et al. 2012). Estimates for total C stocks (biomass and SOC in 0–1 m depth) in tropical forests roughly vary between 400 and 800 Pg. It is estimated that total C storage amounts to 55 percent of global forest C stocks, with 56–58 percent of C stored in living biomass, 32–41 percent in soils, and 12 percent in dead wood and litter (Pan et al. 2011; Meister et al. 2012), although Jobbagy and Jackson (2000) and Sabine et al. (2004) reported an inverse relation between C stored in vegetation (44 percent) and soils within a depth of 1 m (56 percent) (Table 3). Tropical evergreen and deciduous forests soils store 474 and 218 Pg C (0–3 m depth), respectively, and more than 10 percent of the terrestrial global C stocks could be found between 1–3 m depth in tropical forest soils (Jobbagy and Jackson 2000).

Large uncertainties about C fluxes remain, as a result of imprecise information on forest cover, inconsistent methods for biomass estimates, or difficulties in measuring belowground C pools (Meister et al. 2012). Only recently more accurate, spatially explicit global-scale estimates for forest biomass and C density are becoming available, due to advances in image processing and accessibility of remote sensing data and an expanding network of forest monitoring plots (Baccini et al. 2012; Houghton et al. 2012; Pan et al. 2013; Anderson-Teixeira et al. 2015). In terms of AGC, limited information exists on the importance of coarse woody debris (> 10 cm in diameter), although this component may represent a third of the C stocks found in AGB in tropical humid forests (Clark et al. 2002) and could even be more important in dry or semi-evergreen tropical forests that are exposed to natural or human disturbances (Meister et al. 2012). Belowground C stocks and processes, such as root biomass and turnover rates remain poorly studied. Robinson (2007) argued that root biomass might be 66 percent above previous estimates, which would mean an additional 49 Pg C stored below tropical forests globally. The spatial distribution of belowground biomass is highly variable. Meister et al. (2012) suggested that higher fine root biomass may be associated with low availability of soil nutrients and water and that fine root turnover is increased by high fluctuations in soil moisture.

Tropical forests have the highest average C density of all forest biomes with an average C density of 282.5 Mg/ha for intact forests and 241.6 Mg/ha for all tropical forests, including regrowth (Pan et al. 2011). Carbon densities vary considerably between different forest types. For instance, Donato et al. (2011) reported that mangrove forests in the Indo-Pacific region store over 1000 Mg C/ha with 49–98 percent of the C located in soils to 3 m depth.

Tropical forests are the most productive biome on earth, comprising between 30 and 40 percent of global terrestrial GPP (40.8 Pg C/yr)

(Beer et al. 2010). This value corresponds approximately to an average of 23 Mg C/ha/yr, which is in accordance with GPP measurements from 15 sites across the tropics that ranged between 19.6–38.1 Mg C/ha/yr (Mahli 2012). An average GPP of 36 Mg C/ha/yr was reported based on data from 29 humid tropical sites (Luyssaert et al. 2007). Inter-annual GPP variability is largely controlled by decreased productivity and/or increased respiration during El Niño-Southern Oscillation (ENSO) conditions, which mainly affect forests in the Amazon and SE-Asia (Heimann and Reichstein 2008; Meister et al. 2012).

Available total biome NPP estimates vary between 13.7 and 21.9 Pg C/yr (Mahli and Grace 2000; Prentice et al. 2001; Saugier et al. 2001; Sabine et al. 2004), representing a quarter to one third of global terrestrial NPP. NPP is extremely variable across tropical forests, reflecting differences in climate, topography, and soil conditions. Estimates compiled by Clark et al. (2001) ranged between 1.7 and 21.7 Mg C/ha/yr across 39 sites. Another set of NPP measurements from 10 sites ranged from 7.3 and 14.4 Mg C/ha/yr (Mahli 2012). Luyssaert et al. (2007) reported an average of 8.6 Mg C/ha/yr for the humid tropics; this value is relatively low in comparison to GPP, due to a high average R_a loss of 23 Mg C/ha/yr, or 63 percent of the mean GPP.

Across the humid tropics most of NPP (37 percent) is allocated to foliage, followed by roots (38 percent), and only about one quarter to the production of wood (Luyssaert et al. 2007). Based on data from over 100 tropical forest sites, Mahli et al. (2011) estimated that 39 percent of NPP is allocated to wood production, 34 percent to foliage, flowers, and fruits, and only 27 percent to fine roots. Overall, information on belowground NPP in tropical forests is scarce and highly variable (Lorenz and Lal 2010). Estimates range between 4.6 and 8.8 Mg C/ha/yr for forest R_h, and range from 3.6–4.1 Mg C/ha/yr for NEP (Lorenz and Lal 2010).

Whether tropical forests currently represent a C source or a sink remains debated. Deforestation is the most important disruption of tropical forest C fluxes, and worldwide LUC emissions result almost entirely from tropical deforestation (Lorenz and Lal 2010; Mahli 2010). Historically, the biome has lost about a quarter of its original biomass, considerably less than boreal or temperate forests (Pan et al. 2013). According to the FAO (2012) tropical deforestation peaked between the 1950s and 70s and has slowed since then. A total net loss of 8 percent of tropical forest cover (1,736,000 km^2) occurred between 1990 and 2007, although net deforestation rates slowed down from 114,000 km^2/yr to 85,000 km^2/yr during this period (Pan et al. 2011). Another recent study based on high-resolution satellite imagery (30 x 30 m) found an annual net loss of almost 72,000 km^2/yr with net deforestation increasing by over 2000 km^2/yr between 2000 and 2012 (Hansen et al. 2013).

Similarly, Kim et al. (2015) reported that deforestation in the humid tropics increased by 62 percent from 2000 to 2010, in comparison to the 1990s, with a peak occurring around 2005. It is important, to keep in mind that forest definitions (e.g., percentage tree cover) and survey methods between different sources, such as the FAO Global Forest Resources Assessment Report and studies based primarily on remote sensing data, are not directly comparable (Kim et al. 2015).

Regardless of net deforestation rates, tropical forests may have been a net C sink of 1.3 Pg/yr during recent decades (Lewis et al. 2009) or at least have a neutral net impact on the global C balance (Mahli 2010). More recent estimates indicate gross tropical deforestation emissions between 2.2 and 2.9 Pg C/year (up to 40 percent of anthropogenic fossil fuel emissions), which are only partially compensated by tropical intact forests and a growing extent of secondary forests, resulting in a net C source of 1.0–1.1 Pg C/yr (Pan et al. 2011; Baccini et al. 2012). Tropical soils alone may have released 1.9 Pg C in the period 2000–2007, and over 6 Pg C since 1990, due to forest conversion (Pan et al. 2011). Harris et al. (2012) estimated gross deforestation emissions of 0.57–1.22 Pg C/yr, without considering changes in SOC.

Brazil and Indonesia, which contain more than a third of tropical forest AGC stocks, roughly accounted for 60 percent of the deforested area in the humid tropics and were responsible for 55 percent of global deforestation emissions between 2000 and 2005 (Hansen et al. 2008; Baccini et al. 2012; Harris et al. 2012). The reduction in deforestation achieved by Brazil (from around 40,000 km^2/yr to 20,000 km^2/yr) between 2003 and 2011 was offset by doubling deforestation rates from under 10,000 km^2/yr to more than 20,000 km^2/yr in Indonesia and by smaller forest cover losses in moist and dry forests of Africa, South America, and Asia (Hansen et al. 2013). Deforestation drivers include increasing demands for beef, soy beans (Brazil), and palm oil (mainly South East Asia) (Lorenz and Lal 2010). Interestingly, Harris et al. (2012) found that 40 percent of the deforestation happened in the dry tropics. The highest amount from global deforestation emission is generated in Latin America (41–54 percent) and in Southeast Asia (32–47 percent, with a large amount resulting from peatland fires and drainage). Africa only contributes between 11 and 14 percent (Mahli 2010; Harris et al. 2012).

It has been estimated that each Mg of C released from deforestation is associated with additional emissions of 0.6 Mg from subsequent degradation of remaining forests (Houghton 1991; Lorenz and Lal 2010; Meister et al. 2012). As much as 8.5 million km^2 of tropical forests may already be degraded (Pan et al. 2013), an area that represents between 40 and 60 percent of the total forested area (Table 3).

Soil organic carbon losses after forest conversion may occur due to enhanced decomposition or top soil erosion, among other factors. Don et al. (2011) reported that the conversion of primary tropical forests to cropland caused SOC losses between 25 and 30 percent, and conversion to grassland reduced SOC by 12 percent (average sampling depth of 32 cm). High SOC losses can be expected from conversion of forests with deep and C rich soils such as peat forests or mangroves. Cultivated peatlands in Indonesia release 10–15 Mg C/ha/yr and global emissions from peatland are estimated to amount to 0.30 Pg/yr (Inubushi et al. 2003; Reijnders and Huijbregts 2008; Van der Werf et al. 2009). Mangrove deforestation may add another 0.02–0.12 Pg C/yr, largely due to release of SOC (Donato et al. 2011). The FAO (2010) reported that mangroves have decreased from 16.1 to 15.6 million hectares since 1990.

Forest Carbon and Global Change

Natural disturbances such as wildfires, storms, and insects can cause large shifts in C cycles on regional and global scales. In addition, human activities are affecting the C cycle in unprecedented ways by altering the natural ecosystem processes that regulate C uptake, storage, and emissions (e.g., Chapin et al. 2009; IPCC 2014; Le Quéré et al. 2014). In the following we elaborate on a number of global perturbations which directly and indirectly influence global C sinks and sources:

1. Increasing atmospheric CO_2 concentrations
2. Climatic changes
3. Nitrogen deposition
4. Loss of biodiversity

Increasing Atmospheric CO_2 Concentrations

Several vegetation process models predict strong positive effects of higher atmospheric CO_2 concentrations ("CO_2 fertilization effect") on forest NPP (Cramer et al. 2001; Cox et al. 2004; Berthelot et al. 2005; Denman et al. 2007; Bonan 2008). However, many models have not included feedback from the nitrogen (N) cycle. Recent findings indicate that limited N availability would reduce the CO_2 fertilization effect on trees (Norby et al. 2010; Piao et al. 2013). Free-air CO_2 enrichment experiments (FACE) in temperate forests showed that a 50 percent increase in atmospheric CO_2 concentrations enhanced NPP by 23 percent over the first years. However, forest growth declined to 9 percent after 8 years of continued CO_2 enrichment as N availability declined over time and limited tree growth (Norby et al. 2010). Using dynamic global vegetation models and experimental data to assess the sensitivity of plant productivity to changes in atmospheric CO_2 concentrations and climate showed an average increase of 7 percent in global NPP as a response to a 48 ppm increase of CO_2 during the past three decades (Piao et al. 2013). For boreal regions CO_2 sensitivity decreased by almost half in models that considered N limitation, in comparison with models that ignored interactions with N (from 21 ± 7 percent GPP increase per 100 ppm CO_2 to 12 ± 8 percent, respectively) (Piao et al. 2013).

Aside from N limitations, increasing temperatures and limited water availability can also reduce the CO_2 fertilization effect, especially in the tropics (Heimann and Reichstein 2008; Arora and Boer 2014). On the other hand drier sites in temperate forests could gain additional benefits from CO_2 fertilization, because it may allow for more effective moisture regulation by leaf stomata (Piao et al. 2013). The comparison of data on maximum growth rates and maximum longevity for 141 temperate tree species suggests that any CO_2 fertilization effect is likely to be offset by an associated reduction in the longevity of forest trees (Bugmann and Bigler 2011).

Increases in AGB reported for a series of Amazonian forest plots over the last decades have been explained by CO_2 fertilization (Phillips et al. 2008). An increase has also been reported for African rainforests (Lewis et al. 2009). However, other studies have found no biomass changes in recent decades (Laurance et al. 2004; Chave et al. 2008). In addition, liana abundance has increased significantly in the Amazon, possibly driven by increased tree mortality and CO_2 levels. These changes in forest structure and composition could decrease C storage in tropical forests (Laurance et al. 2014).

Climatic Changes

While climate events can damage forests, and thus affect the C cycle in many ways, the emphasis here is on changes driven by drought and increased temperatures. Observed warming has been most extreme at higher latitudes (Pan et al. 2013). Forest productivity in the northern boreal zone is limited by low temperatures, and often by low nutrient availability. In the boreal region, higher air temperature will extend the growing season and thereby increase production. However, dendrochronological studies have shown site-specific growth declines which were explained by drought stress or nutrient limitation (Silva et al. 2010). A recent decrease in Alaskan forest production due to warmer and drier conditions (Beck et al. 2011) suggests that climate change is also affecting C input in northern ecosystems. Furthermore, warming of surface soils and melting of permafrost soils may lead to large increases in C emissions at northern latitudes (Conant et al. 2008; Billings et al. 2012). The decline in sink strength in high latitude forests has been explained by reduced C inputs and accelerated soil organic matter decomposition, which releases previously stabilized SOC (Hayes et al. 2011). Temperature increases may also release large amounts of CH_4 from deposits of hydrates covered by permafrost (Mascarelli 2009). Changes in temperature and rainfall patterns may increase the occurrence of forest fires (Nabuurs et al. 2007; Pan et al. 2013) and could switch the boreal biome from a sink to a net source of C (Hayes et al. 2011). Increased winter minimum and year-round temperatures have been linked to recent large-scale bark beetle outbreaks (Breshears et al. 2005; Bentz et al. 2010). As mentioned above, insect outbreaks reduce photosynthetic C uptake through defoliation and enhance R_e by insect respiration and accelerated decomposition (Kurz et al. 2008; Edburg et al. 2012).

The effects of increased temperatures on soil C fluxes are not well understood. Global soil respiration (R_a by roots and R_h by soil organisms) amounted to 98 ± 12 Pg C/year in 2008 (85 Pg C excluding agricultural soils), with almost 70 percent originating in the tropics (Bond-Lamberty and Thomson 2010). Global soil respiration increased by 0.1 Pg C annually between 1989 and 2008 with the highest relative increase occurring in boreal regions (Bond-Lamberty and Thomson 2010). Recent data from forest plots all over the world indicate that increased soil respiration is primarily the result of higher soil C supplies from increased global GPP (Bond-Lamberty and Thomson 2010; Chen et al. 2014).

An increase in temperature may be beneficial for tree growth at temperate sites as long as the water supply is sufficient. In areas, where production is limited by water availability, growth and yield are predicted to decrease due to climate change (Resco de Dios et al. 2007; Arora and Boer 2014). The observation that climate-induced tree mortality is occurring not only in semi-arid regions but also in mesic forests suggests that the global rise in temperature may be a common driver affecting local, regional, and global C budgets (Breshears and Allen 2002; Adams et al. 2009; van Mantgem et al. 2009). Models predict that C storage in the tropics will decline with warming as a result of increased R_h (Cramer et al. 2001), increased R_a (White et al. 2000), decreased NPP (White et al. 2000; Cramer et al. 2001), and/or forest dieback (White et al. 2000). Plot scale measurements showed that severe drought stress led to major reductions in forest productivity in a multiyear forest drought experiment in Brazilian Amazonia (Brando et al. 2008). Long-term observations from a wet tropical rain forest in Costa Rica showed significant negative impacts on wood productivity under modest levels of warming and drying (Clark et al. 2010). Elevated tree mortality and lowered tree growth have been reported from all tropical regions during extreme ENSO events (Clark 2004; McDowell et al. 2008; Arora and Boer 2014).

Based on models that incorporate changes in CO_2 levels, temperature and precipitation, terrestrial ecosystems will remain a C sink during the 21st century under all emission scenarios used by the IPCC (see Moss et al. 2010), despite differences between biomes (Arora and Boer 2014). Whereas mid to high latitudes will increase C storage, tropical forests may be emission sources or remain C neutral. The Amazon basin may lose about 20 Pg C between 2006 and 2100, independent from the emission scenario, due to decreased precipitation (Arora and Boer 2014). A recent analysis of data on biomass dynamics from an extensive network of plots showed that the net C-sink in the Amazon basin has decreased by 30 percent (to 0.38 Pg/yr) in the 2000s in comparison to the previous decade due to increased tree mortality and relatively constant levels of productivity. Observed changes in tree mortality may be associated with increased climate variability and accelerated forest dynamics (Brienen et al. 2015). On the other hand, Huntingford et al. (2013) argued that the tropical biome might be more resilient than previously thought, based on 22 models that incorporated changes in CO_2 levels, temperature and precipitation. Tropical forests could act as a sink throughout the 21st century, although C storage in living biomass is likely to peak at the end of this period (Huntingford et al. 2013). Overall uncertainties about the impacts of increased CO_2 levels and climate change on the C balance of tropical forests remain very high (Wood et al. 2012).

Nitrogen Deposition

The global N cycle has been strongly altered by human activity over the past century (Galloway et al. 2008). N is a limiting nutrient in many ecosystems (Vitousek and Howarth 1991). Increases of available N due to agriculture and NO_x pollution can alter NPP and nutrient turnover rate through decomposition of organic matter (e.g., Kaspari et al. 2008). In forests with low N supplies, higher rates of N deposition may increase foliar biomass production and C allocation to foliage (Hyvönen et al. 2007; Litton et al. 2007). This may lead to greater photosynthesis per unit area of forest (Högberg 1997). Additional N deposition has been responsible for about 10 percent of the total C sequestration in European forests during the second half of the 20th century and could increase tree C storage by 0.31 Pg/yr on a global scale (De Vries et al. 2006; Magnani et al. 2007; Thomas et al. 2010). However, results vary among ecosystems as not all tree species respond to N deposition with increased growth (Thomas et al. 2010). Xia and Wan (2008) found that broad leaved tree biomass responds more positively to N addition than the biomass of coniferous trees. Furthermore,

N deposition may enhance C uptake only in young forests (Churkina et al. 2007). Increased N deposition may also result in C loss due to enhanced decomposition rates of high-quality litter (low C:N ratio) (Hyvönen et al. 2007). Overall, the magnitude of the impacts of N deposition on the C cycle through N fertilization and saturation is difficult to quantify, especially in the long term (Tyrrell et al. 2012).

Loss of Biodiversity

Species extinction is currently exceeding natural rates between 100 and 1000 times, due to the expansion of agriculture and urban areas, human induced disturbances, and the spread of invasive species. At the same time we are learning that biodiversity is critical for sustaining ecosystem functions (Rockström et al. 2009; Thompson et al. 2012). Consistent positive relationships exist between biological diversity and C storage on a global scale (Strassburg et al. 2010). Consequently, species loss may impact global C cycles. Understanding the relationships between diversity and C storage is critical for the success of forest based GHG mitigation strategies, such as REDD (Midgley et al. 2010; Strassburg et al. 2010; Thompson et al. 2012).

Species diversity and the composition of specific functional traits may affect soil R_h rates or the allocation of plant resources to different C pools, and thus influence forest NEP (e.g., Catovsky et al. 2002; De Deyn et al. 2008). Species complementarity can increase ecosystem productivity by optimizing resource use. Examples of complementarity include differing levels of shade tolerance or rooting depths, facilitation of increased productivity by nitrogen fixers or litter quality, and increased resilience against pests and disturbances (Catovsky et al. 2002; Kirby and Potvin 2007; Larjavaara 2008; Carroll et al. 2012; Thompson et al. 2012).

The highest levels of biological diversity and C storage converge in tropical forests, which are also facing the highest rates of deforestation and degradation (e.g., Strassburg et al. 2010). Cavanaugh et al. (2014) found that taxonomic diversity and the dominance of genera with very large trees were positively correlated to C storage across the tropics. Modeling the impact of different functional trait-based extinction scenarios on the AGC storage of a tropical forest in Panama showed that the removal of trees for commercial exploitation (high wood density and large wood volume) may result in an AGC loss of up to 70 percent (Bunker et al. 2005). Increased CO_2 levels and disturbance rates might favor fast growing species, resulting in AGC reductions of up to 34 percent, whereas decreased precipitation may induce a shift to drought tolerant species which store more AGC (+10 percent) (Bunker et al. 2005). In addition, extinctions decrease the variation of functional traits, which in turn leads to stronger impacts of further species losses on C storage (loss of biological insurance) (Bunker et al. 2005).

Forest Carbon Management and Policy

Global Policies

Forest management and conservation have important effects on C density and fluxes. Fifty-four percent of the world's forests are currently managed for wood and non-wood products, or the provision of ecosystem services. Thirteen percent of the global forest cover is located within protected areas (FAO 2010). Central questions for forest C management are: What do we currently know about management options to increase C sequestration and storage? Can forestry help to compensate a potential decline of forest C sinks while maximizing associated economic, social, and environmental benefits from forests?

Forest management for C sequestration has received growing attention after the inclusion of forestry activities in the UNFCCC Kyoto Protocol, which allowed developed countries to use GHG sinks from LUC (afforestation and reforestation)[2] to meet their emission reduction commitments. However, the role of forests in GHG reduction initiatives, such as Clean Development Mechanisms (CDM) remained marginal and controversial, due to complex rules and limited value of temporary carbon credits obtainable from forestry activities (Fahey et al. 2010; Kumar and Nair 2011; Logan-Hines et al. 2012; Wollenberg et al. 2012). Since the conclusion of its first commitment period 2008–2012, the Kyoto Protocol is only upheld by a decreased number of countries, which represent merely 15 percent of global CO_2 emissions (UNFCCC 2012; WRI 2012).

Management and conservation of existing forests were originally excluded from eligible mitigation activities by the UNFCCC (UNFCCC 2002). However, the Conference of the Parties (COP) 13 in Bali 2007 agreed that reducing emissions from deforestation and forest degradation (REDD) in developing countries should be incorporated into a post-Kyoto framework, as a cost-effective strategy to mitigate GHG, while potentially providing

---

[2] Different definitions exist for afforestation and reforestation. Afforestation usually refers to the establishment of forest on land that has not been forested for a defined period, e.g., 20 to 50 years (Watson et al. 2000). According to FAO (2010) afforestation refers to planting trees on previously non-forested land (increases forest area). Reforestation is defined as replanting or natural regeneration of recently cut down forest (no change in area).

additional benefits such as biodiversity conservation and poverty alleviation (Logan-Hines et al. 2012). Eliminating worldwide deforestation by 2030 could result in a GHG mitigation potential between 2.3 and 5.8 Pg CO_2 eq./yr (Smith et al. 2013). Although practical, legal, and financial aspects of REDD remain undefined on a global scale in the absence of a legally binding international agreement on climate change, this mechanism is becoming a reality on the national and subnational level. Pilot REDD initiatives are carried out voluntarily by developing countries, and mitigation results are required to be fully measured, reported, and verified (MRV). Hundreds of projects are underway in more than 40 tropical countries. The elaboration of 'REDD readiness' plans and REDD projects are currently funded, e.g., by the Forest Carbon Partnership Facility of the World Bank, the UN REDD programme, or bilateral initiatives (Angelsen et al. 2010; Logan-Hines et al. 2012; Sills et al. 2014). During the 2014 United Nation's Climate summit several governments, non-governmental organizations, companies, and indigenous groups endorsed the New York Declaration on Forests which consists of a legally non-binding declaration and action plan to reduce deforestation to 50 percent by 2020 and end deforestation until 2030. The implementation of this plan could mitigate between 4.5 and 8.8 Pg CO_2 annually by 2030 (UN 2014).

Forest Carbon Management

Maximizing forest C storage while simultaneously ensuring other forest functions and services represents a novel challenge for forest managers. The potential of GHG mitigation from forestry may amount to 13.8 Pg C/yr by 2030, although uncertainties about emissions from agriculture, forestry, and other land-use (AFOLU) are much larger than in other sectors (Smith et al. 2014). Forest-based GHG mitigation can be achieved through the following major pathways (Canadell and Raupach 2008; Lorenz and Lal 2010; Carroll et al. 2012; IPCC 2014):

1. Reforestation
2. Increasing stand C density by forest management
3. Reduced emissions from deforestation and forest degradation

The impact of forestry activities on GHG emissions further depends on the type and durability of wood products, the degree to which these products replace GHG-intensive alternatives, such as fossil fuels, steel or concrete, and on GHG emissions throughout the entire life cycle of wood based products, e.g., from harvest, transportation, and processing (e.g., Profft et al. 2009; Fahey et al. 2010); however, a detailed analysis of these aspects is beyond the scope of this chapter.

Afforestation and Reforestation

Planted forests[3] comprise 7 percent (> 2.6 million km^2) of the global forest cover, and 5 percent can be classified as plantations for commercial use (FAO 2010; Pan et al. 2013). Half of the world's planted forests produce industrial round-wood, although the importance of pulp, fiber, and bioenergy plantations is rapidly increasing. Three quarters of all planted forests are located in East Asia, North America, and Europe. Afforestation in the temperate and boreal biomes increased from 74,750 km^2/yr during the 1990s to 94,900 km^2/yr between 2000 and 2010, with more than 40 percent of the annual afforestation carried out by China (Pan et al. 2011). Watson et al. (2000) estimated the C sink from afforestation and reforestation to be 0.03–0.2 Pg/yr for the period between 2008 and 2012 in temperate regions, whereas the estimate for boreal zones is close to zero (estimates exclude SOC and dead organic matter). The afforested area in the tropics increased from 21,200 km^2/yr during the 1990s to 35,540 km^2/yr (Pan et al. 2011). Watson et al. (2000) estimated a C sink of 0.2–0.4 Pg/yr in the tropics for the period between 2008 and 2012, assuming a reforested area of 26,000 km^2/yr. Due to the interdependence between the forestry and agricultural sectors, mitigation measures such as improved land management, waste reduction, and dietary changes by consumers could result in spare land for afforestation with a mitigation potential between 6.1 and 16.5 Pg CO_2 eq./yr by 2050 (Smith et al. 2013).

The term forest plantation usually refers to even-aged monoculture stands established by planting (Carle et al. 2002). Plantations often undergo intensive levels of management, such as site preparation, fertilization, and pest control. Genetically improved and uniform varieties are commonly used to maximize timber production (Pregitzer and Euskirchen 2004; Lorenz and Lal 2010). Intensive management may increase C sequestration capacity of plantations, but potential tradeoffs need to be considered. Fertilization stimulates stand productivity, but may also enhance microbial activity (R_h) and induce NO_x and N_2O emissions (Jandl et al. 2007; Sanchez et al. 2007; Carroll et al. 2012). Similarly, site preparation and liming can lead to SOC losses by increasing soil respiration and DOC leaching (Jandl et al. 2007; Liao et al. 2010). During the initial stage of plantation establishment, R_h from disturbed soil and organic matter is usually high, while NPP of the young trees is low. Consequently, forest plantation NEP may remain negative for up to 15 years (Lorenz and Lal 2010).

The impacts of afforestation on SOC fluxes remain widely unclear. Different studies have reported either SOC gains or losses after afforestation

---

[3] Forests predominantly composed of trees established through planting and/or deliberate seeding (FAO 2010).

of previously non-forested land (Jandl et al. 2007). Whereas C may be accumulated on the forest floor and in top-soils, C losses—even net reductions of SOC—have been observed in some cases for deeper soil layers. One explanation is the stimulation of microbial activity by fresh C inputs in the form of plant roots (priming effect), which induces losses of stabilized SOC, due to increased soil respiration and DOC leaching (Jandl et al. 2007; De Deyn et al. 2008; Mobley 2011). Apparently soils of afforested cropland accumulate C quickly, but most of it remains unstable and is easily lost by decomposition. Stable SOC (e.g., complexes of SOC and clay particles) accumulates at a very slow rate of several kg/ha/yr (Jandl et al. 2007). Berthrong et al. (2009) analyzed soil variables from over 150 plantation sites worldwide. They found that SOC content (30 cm depth) decreased by 6.7 percent and soil N decreased by 15 percent as a result of afforestation. Exchangeable cations were significantly depleted on plantations of *Pinus spp.*, *Eucalyptus spp.*, and other conifer species, although losses of soil fertility and SOC stocks may be partially compensated by retaining harvesting residues on the site (Sanchez et al. 2007; Berthrong et al. 2009; Lorenz and Lal 2010).

Around 75 percent of planted forests consist of native species, although some countries, e.g., in Oceania and South America, use mostly exotic tree species (FAO 2010). Worldwide, the most common plantation species are *Pinus spp.* with 20 percent of the afforested area and *Eucalyptus spp.*, accounting for 10 percent (Fig. 2) (Carle et al. 2002; FAO 2010; Lorenz and Lal 2010). Under humid tropical conditions, fast growing species can approximately reach average AGB increment values of 20 Mg/ha/yr (*Pinus spp.*) to 40 Mg/ha/yr (*Eucalyptus spp.*), whereas natural forest regeneration may reach up to 13 Mg/ha/yr AGB growth (Penman et al. 2003).

Species selection and composition play an important role for the C dynamics of forest plantations (Potvin et al. 2011); it was found, for instance, that species richness and the proportion of N-fixing trees were positively correlated with C storage in tree biomass of multi-species plantations in Panama (Ruiz-Jaen and Potvin 2010). Average basal area growth of native timber species from tropical Australia (*Agathis* ssp. and *Eucalyptus spp.*) increased by 55 percent, if these species were grown in mixtures (Erskine et al. 2006).

In terms of SOC storage, Resh et al. (2002) reported that tropical plantations of N-fixing tree species (families Fabaceae–Mimosoidae and Casuarinaceae) sequestered 1.1 Mg/ha/yr, in comparison to plantations of *Eucalyptus spp.*, which effected no changes in SOC after afforestation of agricultural land. Similarly, Sang et al. (2013), found that plantations of *Eucalyptus urophylla* S.T. Blakein Vietnam stored less SOC than plantations of N-fixing *Acacia mangium* Willd. or adjacent secondary forests; they further

Fig. 2. Eighteen year-old plantation of *Eucalyptus saligna* Sm. for pulp and sawn wood production in Paraná, Brazil (Photo credit: Przemyslaw Walotek).

predicted that repeated rotations (< 10 years) of *Eucalyptus urophylla* would reduce SOC significantly. According to Russell et al. (2004), SOC storage increased significantly in plantations of *Cedrela odorata* L. and *Cordia alliodora* Cham. in Costa Rica, if palms and understory monocots were included.

Agroforestry, the combination of crops or livestock with trees, is another way to increase tree cover and soil C storage in altered landscapes. Worldwide there are more than 10 million km² of agricultural land with a tree cover of more than 10 percent, which directly provide livelihoods for over 550 million people (FAO 2012; Kapos et al. 2012). Around 6.3 million km² are potentially available for the adoption of agroforestry, resulting in a total mitigation potential of nearly 0.6 Pg C/yr by 2040 (Jose and Bardhan 2012). Agroforestry is recognized as a climate change mitigation strategy by the UNFCCC and may play an important role in the implementation of mechanisms such as REDD and nationally appropriate mitigation actions (NAMAs) in developing countries (Jose and Bardhan 2012; Kapos et al. 2012; Harvey et al. 2014).

In the tropics, where deforestation and forest degradation represent the most important disruption of the C cycle, reforestation, forest restoration, and deforestation avoidance (e.g., conservation, improvement of agricultural management) are by far the most important pathways for GHG mitigation. In higher latitudes forest management that aims to increase

C density has the highest GHG mitigation potential among forestry based activities (Smith et al. 2014).

Management of Carbon Density in Temperate and Boreal Forests

In Central Europe forestry has aimed to sustain timber yields for centuries. The current positive C balance of boreal and temperate forests is mainly the result of reforestation and afforestation, motivated by demand for wood (Pan et al. 2011). In addition, C density per unit area (Mg/ha) is affected by a variety of management activities. For instance, wood harvest inherently represents a C export from forests (Jandl et al. 2007), although a variety of management options may minimize C losses from biomass and soils during and after harvesting activities.

Rotation cycles in managed temperate and boreal forests typically range from 25 to 150 years, whereas natural succession cycles may continue for centuries. Short rotations maximize C sequestration rates, but C storage over multiple harvesting cycles may be reduced to one third or less, compared to late successional forests. Longer cycles, in turn, maximize C storage and NEP on a landscape level (Luyssaert et al. 2008; Lorenz and Lal 2010; Carroll et al. 2012). In the case of even aged stands, AGC sequestration and yields over time can be maximized by harvesting when periodic annual increment (volume added/ha/yr) drops to the amount of mean annual increment (stand volume divided by age). Rotation cycles beyond this point will lead to decreasing harvestable volume over time; however, shorter rotation cycles result in higher fossil fuel emissions (Carroll et al. 2012). Longer rotations also stabilize SOC by decreasing soil disturbance (Jandl et al. 2007).

Disturbance is also reduced by low impact harvesting which aims to minimize the damage to residual vegetation, soil compaction, and erosion (Lorenz and Lal 2010). Nabuurs et al. (2013) recommended the conservation of forests on sensitive sites (high SOC stocks, steep slopes) and the implementation of continuous-cover management where it is possible. Clear cutting, on the other hand, may result in sustained SOC and litter C losses for decades, although evidence is not conclusive for all forest types, and it may take several rotations (i.e., centuries), to significantly affect SOC stocks (Jandl et al. 2007; Lorenz and Lal 2010; Carroll et al. 2012). Leaving harvesting residuals on site may increase SOC accumulation, although other variables, such as soil clay content play an important role in this context (Lorenz and Lal 2010). Overall forest SOC seems to be fairly resistant to silvicultural treatments, as long as soils are not heavily disturbed and forests are not converted to other land-use types (Fahey et al. 2010; Carroll et al. 2012).

Thinning is another management activity that affects forest C fluxes, although knowledge on optimum thinning levels to maximize C storage remains scarce. Thinning immediately reduces C stocks due to wood removal, decreased litter input and the decomposition of organic matter. On the other hand, thinning may stimulate photosynthesis in the understory, increase stand resilience against disturbances, decrease mortality, and thus ensure sustained productivity in the long term. The net effects of thinning on SOC pools do not appear to be significant (Jandl et al. 2007; Lorenz and Lal 2010). Carroll et al. (2012) argued that thinning cannot increase productivity above a limit imposed by the physical environment. Thus, in the best case thinning is C neutral, unless C storage in wood products from thinning operations is taken into account.

Fire and Extreme Events

Fire is a major element of C cycling in many boreal and temperate forests. Management responses include fire suppression or thinning and fuel removal (e.g., prescribed burning). Drawbacks of fire suppression, especially in fire adapted forests, include fuel build-up, insect outbreaks, a reduced proportion of large trees, and uncertain shifts in species composition. Adaptive strategies play an important role in successful forest fire management. For instance, droughts, in combination with increased fuel loads from fire suppression, have increased the risk of destructive forest fires in the Western United States. Prescribed burning, on the other hand, decreases C stocks in the short term, but increases resilience against catastrophic fires, which may cause continued GHG emissions for decades, e.g., from soil erosion and decomposing organic matter. Under natural conditions, unmanaged fire-dependent forest ecosystems appear to be net C sinks (Canadell and Raupach 2008; Lorenz and Lal 2010; Carroll et al. 2012).

Building stand resilience to disturbance from fire, insects, and extreme weather events is fundamental for C management in the face of uncertainties of global change. Management activities that increase resilience include the establishment of mixed species stands, the incorporation of different age classes (e.g., by under-planting or supporting natural regeneration) and the careful selection of adapted species, genotypes, and seed provenances. In this context it is important to consider potential species distribution shifts (Canadell and Raupach 2008; Larjavaara 2008; Lorenz and Lal 2010; Carroll et al. 2012).

Species Management

The role of functional diversity on ecosystem productivity and its implications for forest management remain debated, as it is difficult to untangle the effects of tree diversity and composition from the influence of physical site conditions, land use history, or environmental change (Jandl et al. 2007; Lorenz and Lal 2010; Dean et al. 2012). One of the most studied mixtures in managed forests is the combination of European beech (*Fagus sylvatica* L.) and Norway spruce (*Picea abies* (L.) Karst.). Pretzsch and Schütze (2009) found that species complementarity increased aboveground productivity up to almost 60 percent in mixed stands of beech and Norway spruce in Southern Germany. With regards to maximizing SOC storage, the inclusion of deep rooting trees is particularly important to transfer C into more stable pools in the mineral soil via root exudates and dead organic matter (De Deyn et al. 2008; Lorenz and Lal 2010).

On the other hand, mixed forests are more complicated to manage, as information on silviculture for diverse stands is lacking in many parts of the world (Larjavaara 2008). Nevertheless, it can be concluded that increased productivity and resilience make certain tree species combinations a favorable option for C management (Jandl et al. 2007), given that current and projected physical site conditions are appropriate for these species.

Management of Carbon Density in Tropical Forests

Roughly 20 percent of tropical forests are subjected to commercial timber extraction. These forests are usually exploited rather than managed. Only 7 percent of tropical timber comes from sustainably managed forests. Reasons for this situation include short term logging concessions (10–60 years), land tenure insecurity, poor regulations and enforcement, and lack of silvicultural knowledge. Moreover, sustainable logging is widely perceived by the timber industry to decrease profits (Canadell and Raupach 2008; Putz et al. 2008a; Putz et al. 2008b; TFF 2009; Blaser et al. 2011; Del Cid-Liccardi et al. 2012).

Understanding the impact of conventional logging practices on C fluxes is critical for quantifying emission reductions achieved by implementing sustainable forest management (i.e., to demonstrate C sequestration additionality with reference to a business as usual scenario) (Del Cid-Liccardi et al. 2012). Rotation cycles in diameter-limit selective logging operations are commonly 30 years or less, which leads to structural and functional forest degradation (Putz et al. 2008a; Blanc et al. 2009; Del Cid-Liccardi et al. 2012; Kapos et al. 2012). Although selective logging removes only a small proportion of stems, commercial timber species commonly store high amounts of C per tree, and their removal may decrease AGC stocks by up to 70 percent (Bunker et al. 2005; Kirby and Potvin 2007). Every logged tree damages up to 20 additional ones (Putz et al. 2008a), and logging roads allow for subsequent degradation and clearing of remaining forest by illegal wood extraction and land colonization (Blanc et al. 2009). Using high resolution LiDAR to estimate the impact of selective logging on AGC stocks in the Peruvian Amazon, Asner et al. (2010) revealed that degraded forests stored 70 percent less AGC, and that forest degradation increased C emissions by 47 percent, compared to emissions from regional deforestation alone. Houghton et al. (2012) reported that C emissions from tropical forest degradation by logging and shifting cultivation add 25–35 percent to the net emissions from land use change.

The thinning of non-commercial species may decrease C stocks even further. Girdling non-timber trees in the Amazon led to doubled growth rates in timber species, but average net C flux from this treatment was approximately four times higher than from conventional selective logging (Blanc et al. 2009).

On the other hand, improved logging practices may decrease C emissions by 30 percent, adding up to a global mitigation potential of 0.16 Pg C/yr, which represents 10–15 percent of emissions from tropical deforestation (Putz et al. 2008a). Reduced impact logging (RIL) aims to minimize impacts from selective logging by a combination of education, planning, enforcing environmental and social standards, appropriate harvest practices, and post-harvest monitoring (Putz et al. 2008b; TFF 2009). Careful planning and harvesting practices, implemented by trained workers, reduce soil damage from skidding and machinery. Directional felling reduces the damage of forest regeneration, improves the silvicultural value for future rotation cycles, and maintains higher C stocks and sequestration potential in the long term. At the same time, RIL reduces timber waste (Putz et al. 2008b). The detection of hollow trees prior to logging may further reduce emissions, as about a quarter of felled trees are not extracted due to hollowness (Putz 2013; Griscom et al. 2014). Long term management plans, moderate logging intensities (e.g., less than 10 stems/ha), and extended harvesting cycles (40–100 years) are essential for ensuring sustained timber yields and reduced C emissions (Blanc et al. 2009; Mazzei et al. 2010; Del Cid-Liccardi et al. 2012). Where natural regeneration of commercial species is not sufficient, reproduction needs to be managed by retaining seed trees or shelter woods, or by enrichment planting (Putz et al. 2008b; Del Cid-Liccardi et al. 2012). More research is needed on species-specific life spans, growth dynamics, and C storage capacities, in order to better plan for sustained timber yields and C stocks on the landscape scale (Kirby and Potvin 2007).

Overall, reconciling environmental sustainability of industrial logging operations with economic viability remains challenging, given that governance in the forestry sector in tropical countries is often weak, and incentives for converting logged over forests into agriculture are strong (Kormos and Zimmermann 2014).

Previously degraded forests and regrowth is more susceptible to anthropogenic fires, which are mostly associated with agricultural activities. Emissions from deforestation and forest degradation by fires in the tropics amount to 1.39 Pg CO_2-eq./yr. For this reason, forestry-based GHG mitigation measures and mechanisms, such as REDD, should also aim to improve agricultural practices near forested areas in the tropics (Kapos et al. 2012; Smith et al. 2014).

Comparison between Carbon Storage in Plantations, Managed, and Unmanaged Forests

Undisturbed old-growth forests represent vast C pools (Luyssaert et al. 2008) that exceed the C storage potentials of other land use options. Forest plantations and managed forests in the temperate and boreal zones store 40–62 percent less AGC and 26–42 less SOC than old-growth forests (Carroll et al. 2012). Tropical forest plantation on average accumulate only 57 percent of the AGB found in natural forests, although biomass growth rates are over twice as high on plantations (Aalde et al. 2006). Average C stored in the biomass of plantations over time is inherently lower than in natural forests, because periodical harvest is usually carried out near the peak of stand volume increment. For instance, Kraenzel et al. (2003) estimated an average 120 Mg/ha AGC stored in Panamanian teak (*Tectona grandis* L.f.) plantation at a harvest age of 20 years; however, the average C storage over several rotations would reduce to only about 60 percent of this amount.

No consistent trend for differences in SOC storage was found between plantations and natural forests, although negative impacts on SOC may result from soil disturbance by management activities (Lorenz and Lal 2010). Comparing C stocks between natural forests and plantations (average 30 years) from 86 paired studies showed that total C (soil and biomass) was on average 28 percent lower (205 Mg C/ha) in the plantations than in natural forests (284 Mg C/ha) across all forest biomes (Liao et al. 2010). Soil macronutrients (N, P, K) decreased on average between 20 and 26 percent which may further reduce NPP rates and induce a negative feedback for C sequestration on plantations (Liao et al. 2010). Furthermore, CH_4 uptake by plantation soils is 80 percent lower than in natural forests (Liao et al. 2010).

Whereas the importance of unmanaged old-growth forests for C storage is evident, their role as a C sink remains debated. Classic concepts of ecosystem dynamics predict a steady state at late successional stages, where GPP declines and approximates R_e (e.g., Odum 1969). Observations from even-aged, mono-species plantations confirm the hypothesis that older stands become C neutral; however, many unmanaged boreal and temperate old-growth forests appear to accumulate AGC and SOC for centuries (e.g., Luyssaert et al. 2008; Lichstein et al. 2009; Schulze et al. 2009). Supported by an extensive database from flux-tower measurements, Schulze et al. (2009) estimated an AGC sink of 0.4 Pg/yr for temperate and boreal old-growth forests; GPP remains constant at maximum level in old stands, which leads to continual accumulation of biomass. At the same time, the probability of a catastrophic event increases exponentially with age, resulting in a mosaic of old-growth and recovering stands in unmanaged forests. Reviewing data from over 500 boreal and temperate forest sites Luyssaert et al. (2008) found that R_h/NPP remained constant, independent of stand ages between 15 and 800 years, which indicates that old growth forests continue to function as C sinks. The authors estimated that old-growth forests (> 200 years) in the temperate and boreal biome account for a C sink of 1.3 ± 0.5 Pg/yr in aboveground and belowground pools. Individual large, old trees from more than 400 temperate, tropical and subtropical species were shown to continuously accumulate biomass at very high rates, despite a possible decline of growth per unit leaf area. Large trees not only represent significant C reservoirs, but may also be disproportionately important for C sequestration (Stephenson et al. 2014), although it is important to keep in mind that this study did not focus on entire forest ecosystems.

Despite the uncertainties about the role of old-growth forests as C sinks, it is evident that the conservation of old growth forests is the most effective strategy for maximizing C storage (Luyssaert 2008; Schulze et al. 2009; Carroll et al. 2012). Although timber plantations provide important services compared to non-forested land, they do not seem to offer net benefits for C sequestration or C storage over natural forests, and their role in GHG mitigation frameworks may have been overrated in comparison to unmanaged and old-growth forests (Schulze et al. 2009; Liao et al. 2010; Lorenz and Lal 2010).

Summary

Current Role of Forests in the Global Carbon Cycle

Forests store between 70–80 percent of terrestrial C. On a global scale forests are lost at a rate of > 50,000 km²/yr, and C fluxes from LUC, mostly tropical deforestation, account for 8 percent of anthropogenic C emissions. Despite global net deforestation, forests represent 50–60 percent of the terrestrial net C sink, removing between 1.2–2.4 Pg C/yr from the atmosphere. Boreal and temperate forests are a net C sink, as their area is slightly growing. Tropical forest cover has decreased by around 10 percent since 1990. Estimates of the C balance of tropical forests range from net emissions (source) of 1.1 Pg C/yr to net uptake rates (sink) of 1.3 Pg C/yr, reflecting substantial uncertainties about C fluxes in this biome. Forests and LUC account for the largest uncertainties within the global C balance, due to inconsistent definitions and data sources on forest cover, deforestation and forest degradation, as well as incomplete information on C pools and processes, especially in terms of belowground C stocks and tropical forests. Regional scale estimates of forest AGC are improving fast, as a result of expanding forest monitoring networks and recent advances and broader availability of remote sensing technologies.

The future role of forests as a global C sink depends on a variety of processes and their interactions. Boreal forest productivity may be positively affected by higher temperatures and longer growing seasons, but there is also an increased potential for massive disturbances from insect outbreaks, forest fires, or drought. Melting permafrost soils may release vast amounts of CH_4 and CO_2. These processes could switch the boreal biome from a C sink into a source. The current sink in temperate forest results from the growth of relatively young, expanding forest areas, an effect that may decline in the future. Elevated CO_2 levels and NO_x pollution appear to increase forest productivity, but long term effects of these anthropogenic alterations are difficult to predict. Drought stress and forest fires, as well as increased insect outbreaks, may reduce forest productivity. Furthermore, pressure on forests is increasing from infrastructure development and a growing demand for resources, such as biomass fuels.

Tropical forests cover the largest area of all forested biomes, store 20–40 percent of terrestrial C, and account for the largest gross forest C sink (1.8 Pg C/yr). At the same time, tropical forests face the highest rates of deforestation and degradation. C stocks and fluxes of tropical forests remain poorly understood. The future role of tropical forests depends foremost on LUC dynamics; however, uncertainties persist even about fundamental figures such as deforestation rates in the tropics. Depending on forest cover definitions and survey methods, estimates range from substantial decreases in net deforestation rates to accelerated forest loss during the last decade. Pressure from deforestation drivers such as global demand for sugar cane, soy, or palm oil remains high. C losses from tropical forest degradation may add another 25–35 percent to emissions from deforestation. Degraded forests become more susceptible to encroachment, drought and wild fires, creating a positive emission feedback cycle.

The impacts of climate change on the C balance of tropical forests are difficult to predict. Recent findings range from increasing forest productivity as a result of CO_2-fertilization, to declining productivity and elevated tree mortality during ENSO events. While there is growing evidence for alterations of the C balance by global change, the direction and magnitude of these imbalances remain unclear, as trends are obscured by many confounding variables and non-linear feedback loops. In addition, tropical forests and their potential responses to climate change are highly variable. It is clear, however, that changes in the C balance of tropical forests will strongly affect atmospheric CO_2 levels. Understanding tropical forests' responses to global change is a priority for future research and policies on forest management, conservation, and GHG mitigation.

Forest Carbon Management and Policy

Maximizing forest C storage while simultaneously ensuring other forest functions and services represents a novel challenge for forest managers. Potential synergies and trade-offs between C storage and other ecosystem services must be balanced under changing environmental conditions. Adaptive management strategies are needed to incorporate the growing knowledge on climate change impacts and ecosystem responses into forest C management.

Afforestation and reforestation have remained of marginal importance for accredited C offset projects under the UNFCCC framework. Nevertheless, the extension of planted forests is growing and currently accounts for 7 percent of the forested area worldwide. Plantations usually consist of fast growing species that sequester important amounts of C. On the other hand, afforestation may initially lead to substantial SOC losses and the average AGC storage over multiple rotation cycles is relatively low, in comparison to natural forests. Silvicultural innovations are needed to maximize C storage and other benefits from forest plantations in the future. The management of mixed species plantations may offer many benefits such as increased productivity and resilience to disturbances associated with global change.

In the tropics reforestation, forest restoration and deforestation avoidance are by far the most important forestry based GHG mitigation pathways. REDD has the potential to support these activities in developing countries. Although REDD currently lacks a legal basis on the international level, the mechanism is currently funded, e.g., by the UN REDD Programme and the World Bank and is becoming a reality in many tropical countries.

The lessons learnt by these pilot projects may help to establish REDD as a viable GHG mitigation strategy and help to stop deforestation in the future.

In the case of managed temperate and boreal forests, timber yield and AGC storage can be maximized simultaneously by optimizing the length of rotation cycles, particularly in relatively even-aged stands with few tree species. Long-term research is needed on the impacts of silvicultural activities on SOC storage, including depths of > 1 m. Forest SOC appears to be relatively stable, as long as heavy soil disturbance is avoided, but long term data are lacking.

About 20 percent of tropical forests are harvested for timber, the vast majority being exploited rather than managed. First steps to improve C management include extending rotation cycles, implementing RIL practices and establishing long-term management plans. Structural impediments for these strategies need to be tackled in the respective countries and on a global scale. International support is needed to ensure capacity building and strong institutions on the regional level, and to create incentives for global markets (e.g., rigorous certification systems, forestry based GHG mitigation mechanisms). To be competitive, sustainable forest management must take into account positive and negative externalities regarding C sequestration and other ecosystem services. Innovative forest based GHG reduction mechanisms, such as REDD, need robust baseline scenarios on the emissions from current forest degradation to understand the mitigation potentials (i.e., additionality) of sustainable alternatives. Further research is needed on belowground C storage and processes, including depths > 1 m, and their responses to management activities (e.g., afforestation, soil disturbance) and global change.

Tree species diversity plays a central role for forest C management. There is growing evidence for relationships between tree species richness, composition, and C storage. Diverse forests are more resilient against disturbances, such as fire, insect outbreaks, and meteorological extreme events (biological insurance). Species complementarity may further increase C density and timber yields. A mechanistic understanding of these relationships is lacking, yet critical, for successful forest C management, which may benefit from synergies between biodiversity conservation and GHG mitigation.

Conservation of natural or old-growth forests is clearly the most effective way to secure massive C storage. As natural forests appear to accumulate C for centuries, we need more long-term and global-scale research on the role of old-growth forest as C sinks. Recently, it has been argued that the importance of afforestation as a GHG mitigation strategy has been overstated by international policy frameworks, as there appear to be no net benefits in comparison to C sequestration and storage by natural forests or secondary re-growth. Effective C management and policies must integrate and support afforestation projects, sustainable management, and conservation to balance GHG mitigation goals with maximizing the multiple benefits provided by forests to societies and the environment. Tropical forests should be a priority for forestry based GHG mitigation policies, due to their outstanding importance for the global C balance, and to their sustained loss and degradation.

Acknowledgements

We would like to thank Claire Fox and Joanna Parkman for their comments on an earlier version of this chapter.

References

Aalde, H., P. Gonzalez, M. Gytarsky, T. Krug, W.A. Kurz, S. Ogle, J. Raison, D. Schoene, N.H. Ravindranath, N.G. Elhassan, L.S. Heath, N. Higuchi, S. Kainja, M. Matsumoto, M.J. Sanz Sánchez, Z. Somogyi, J.B. Carle and I.K. Murthy. 2006. Forest land. pp. 4.1–4.83. *In*: S. Eggleston, L. Buendia, K. Miwa, T. Ngara and K. Tanabe (eds.). 2006 IPCC Guidelines for National Greenhouse Gas Inventories. Volume 4. Agriculture, Forestry and Other Land Use. Institute for Global Environmental Strategies, Kanagawa, Japan.
Adams, H.D., M. Guardiola-Claramonte, G.A. Barron-Gafford, J. Camilo Villegas, D.D. Breshears, C.B. Zou, P.A. Troch and T.E. Huxman. 2009. Temperature sensitivity of drought-induced tree mortality portends increased regional die-off under global-change-type drought. Proc. Natl. Acad. Sci. USA 106: 7063–7066.
Anderson-Teixeira, K.J., S.J. Davies, A.C. Bennett, E.B. Gonzalez-Akre, H.C. Muller-Landau, S.J. Wright, K. Abu Salim, A.M. Almeyda Zambrano, A. Alonso, J.L. Baltzer, Y. Basset, N.A. Bourg, E.N. Broadbent, W.Y. Brockelman, S. Bunyavejchewin, D.F.R.P. Burslem, N. Butt, M. Cao, D. Cardenas, G.B. Chuyong, K. Clay, S. Cordell, H.S. Dattaraja, X. Deng, M. Detto, X. Du, A. Duque, D.L. Erikson, C.E.N. Ewango, G.A. Fischer, C. Fletcher, R.B. Foster, C.P. Giardina, G.S. Gilbert, N. Gunatilleke, S. Gunatilleke, Z. Hao, W.W. Hargrove, T.B. Hart, B.C.H. Hau, F. He, F.M. Hoffman, R.W. Howe, S.P. Hubbell, F.M. Inman-Narahari, P.A. Jansen, M. Jiang, D.J. Johnson, M. Kanzaki, A.R. Kassim, D. Kenfack, S. Kibet, M.F. Kinnaird, L. Korte, K. Kral, J. Kumar, A.J. Larson, Y. Li, X. Li, S. Liu, S.K.Y. Lum, J.A. Lutz, K. Ma, D.M. Maddalena, J.-R. Makana, Y. Malhi, T. Marthews, R. Mat Serudin, S.M. McMahon, W.J. McShea, H.R. Memiaghe, X. Mi, T. Mizuno, M. Morecroft, J.A. Myers, V. Novotny, A.A. de Oliveira, P.S. Ong, D.A. Orwig, R. Ostertag, J. den Ouden, G.G. Parker, R.P. Phillips, L. Sack, M.N. Sainge, W. Sang, K. Sri-ngernyuang, R. Sukumar, I. Sun, W. Sungpalee, H. Sathyanarayana Suresh, S. Tan, S.C. Thomas, D.W. Thomas, J. Thompson, B.L. Turner, M. Uriarte, R. Valencia, M.I. Vallejo, A. Vicentini, T. Vrška, X. Wang, X. Wang, G. Weiblen, A. Wolf, H. Xu, S. Yap and J. Zimmerman. 2015. CTFS-Forest GEO: a worldwide network monitoring forests in an era of global change. Glob. Change Biol. 21: 528–549.
Angelsen, A., M. Brockhaus, M. Kanninen, E. Sills, W.D. Sunderlin and S. Wertz-Kanounnikoff. 2010. La Implementación de REDD+: national strategy and policy options. CIFOR, Bogor, Indonesia.
Apps, M.J., W.A. Kurz, R.J. Luxmore, L.O. Nilsson, R.A. Sedjo, R. Schmidt, L.G. Simpson and T.S. Vinson. 1993. Boreal forests and tundra. Water Air Soil Poll. 70: 39–53.

Arora, V.K. and G.J. Boer. 2014. Terrestrial ecosystems response to future changes in climate and atmospheric CO_2 concentration. Biogeosciences 11: 4157–4171.

Asner, G.P., G.V.N. Powell, J. Mascaroa, D.E. Knapp, J.K. Clark, J. Jacobson, T. Kennedy-Bowdoin, A. Balaji, G. Paez-Acosta, E. Victoria, L. Secada, M. Valqui and R.F. Hughes. 2010. High-resolution forest carbon stocks and emissions in the Amazon. Proc. Natl. Acad. Sci. USA 107: 16738–16742.

Baccini, A., S.J. Goetz, W.S. Walker, N.T. Laporte, M. Sun, D. Sulla-Menashe, J. Hackler, P.S.A. Beck, R. Dubayah, M.A. Friedl, S. Samanta and R.A. Houghton. 2012. Estimated carbon dioxide emissions from tropical deforestation improved by carbon-density maps. Nat. Clim. Change 2: 182–185.

Balshi, M.S., A.D. McGuire, Q. Zhuang, J. Melillo, D.W. Kicklighter, E. Kasischke, C. Wirth, M. Flannigan, J. Harden, J.S. Clein, T.J. Burnside, J. McAllister, W.A. Kurz, M. Apps and A. Shvidenko. 2007. The role of historical fire disturbance in the carbon dynamics of the pan-boreal region: a process-based analysis. J. Geophys. Res. 112: G02029.

Beck, P.S.A., G.P. Juday, C. Alix, V.A. Barber, S.E. Winslow, E.E. Sousa, P. Heiser, J.D. Herriges and S.J. Goetz. 2011. Changes in forest productivity across Alaska consistent with biome shift. Ecol. Lett. 14: 373–379.

Beer, C., M. Reichstein, E. Tomelleri, P. Ciais, M. Jung, N. Carvalhais, C. Rödenbeck, M.A. Arain, D. Baldocchi, G.B. Bonan, A. Bondeau, A. Cescatti, G. Lasslop, A. Lindroth, M. Lomas, S. Luyssaert, H. Margolis, K.W. Oleson, O. Roupsard, E. Veenendaal, N. Viovy, C. Williams, F.I. Woodward and D. Papale. 2010. Terrestrial gross carbon dioxide uptake: global distribution and covariation with climate. Science 329: 834–838.

Bentz, B.J., J. Regniere, C.J. Fettig, E.M. Hansen, J.L. Hayes, J.A. Hicke, R.G. Kelsey, J.F. Negron and S.J. Seybold. 2010. Climate change and bark beetles of the western United States and Canada: direct and indirect effects. BioScience 60: 602–613.

Bergeron, Y. and K.A. Harper. 2009. Old-growth forests in the Canadian boreal: the exception rather than the rule? pp. 285–300. *In*: C. Wirth, G. Gleixner and M. Heimann (eds.). Old-Growth Forests: Function, Fate and Value. Ecological Studies 207. Springer, Berlin.

Bernier, P.Y. and M.J. Apps. 2006. Knowledge gaps and challenges in forest ecosystems under climate change: a look at the temperate and boreal forests of North America. pp. 333–353. *In*: J.S. Bhatti, R. Lal, M.J. Apps and M.A. Price (eds.). Climate Change and Managed Ecosystems. Taylor & Francis, Boca Raton, Florida, USA.

Berthelot, M., P. Friedlingstein, P. Ciais, J.L. Dufresne and P. Monfray. 2005. How uncertainties in future climate change predictions translate into future terrestrial carbon fluxes. Glob. Change Biol. 11: 959–970.

Berthrong, S.T., E.G. Jobbágy and R.B. Jackson. 2009. A global meta-analysis of soil exchangeable cations, pH, carbon, and nitrogen with afforestation. Ecol. Appl. 19: 2228–2241.

Billings, S.A., S.E. Ziegler, R. Benner, R. Richter and W.H. Schlesinger. 2012. Predicting carbon cycle feedbacks to climate: integrating the right tools for the job. EOS 93: 188–189.

Black, T.A., D. Gaumont-Guay, R.S. Jassal, B.D. Amiro, P.G. Jarvis, S.T. Gower, F.M. Kelliher, A. Dunn and S.C. Wofsy. 2005. Measurement of CO_2 exchange between boreal forest and the atmosphere. pp. 151–85. *In*: H. Griffiths and P.J. Jarvis (eds.). The Carbon Balance of Forest Biomes. Taylor and Francis Group, New York, USA.

Blanc, L., M. Echard, B. Herault, D. Bonal, E. Marcon, J. Chave and C. Baraloto. 2009. Dynamics of aboveground carbon stocks in a selectively logged tropical forest. Ecol. Appl. 19: 1397–1404.

Blaser, J., A. Sarre, D. Poore and S. Johnson. 2011. Status of Tropical Forest Management 2011. ITTO Technical Series No. 38. International Tropical Timber Organization, Yokohama, Japan.

Bonan, G.B. 2008. Forests and climate change: forcings, feedbacks, and the climate benefits of forests. Science 320: 1444–1449.

Bond-Lamberty, B. and A. Thomson. 2010. Temperature-associated increase in the global soil respiration record. Nature 464: 579–583.

Bond-Lamberty, B., S.D. Peckham, D.E. Ahl and S.T. Gower. 2007. The dominance of fire in determining carbon balance of the central Canadian boreal forest. Nature 450: 89–92.

Brando, P.M., D.C. Nepstad, E.A. Davidson, S.E. Trumbore, D. Ray and P. Camargo. 2008. Drought effects on litterfall, wood production and belowground carbon cycling in an Amazon forest: results of a throughfall reduction experiment. Philos. T. Roy. Soc. B 363: 1839–1848.

Breshears, D.D. and C.D. Allen. 2002. The importance of rapid, disturbance-induced losses in carbon management and sequestration. Ecological sounding. Glob. Ecol. Biogeogr. 11: 1–5.

Breshears, D.D., N.S. Cobb, P.M. Rich, K.P. Price, C.D. Allen, R.G. Balice, W.H. Romme, J. Kastens, M.L. Floyd, J. Belnap, J.J. Anderson, O.B. Myers and C.W. Meyer. 2005. Regional vegetation die-off in response to global-change-type drought. P. Natl. Acad. Sci. USA 102: 15144–15148.

Brienen, R.J.W., O.L. Phillips, T.R. Feldpausch, E. Gloor, T.R. Baker, J. Lloyd, G. Lopez-Gonzalez, A. Monteagudo-Mendoza, Y. Malhi, S.L. Lewis, R. Vásquez Martinez, M. Alexiades, E. Álvarez Dávila, P. Alvarez-Loayza, A. Andrade, L.E.O.C. Aragão, A. Araujo-Murakami, E.J.M.M. Arets, L. Arroyo, G.A. Aymard, O.S. Bánki, C. Baraloto, J. Barroso, D. Bonal, R.G.A. Boot, J.L.C. Camargo, C.V. Castilho, V. Chama, K.J. Chao, J. Chave, J.A. Comiskey, F. Cornejo Valverde, L. da Costa, E.A. de Oliveira, A. Di Fiore, T.L. Erwin, S. Fauset, M. Forsthofer, D.R. Galbraith, E. S. Grahame, N. Groot, B. Hérault, N. Higuchi, E.N. Honorio Coronado, H. Keeling, T.J. Killeen, W.F. Laurance, S. Laurance, J. Licona, W.E. Magnussen, B.S. Marimon, B.H. Marimon-Junior, C. Mendoza, D.A. Neill, E.M. Nogueira, P. Núñez, N.C. Pallqui Camacho, A. Parada, G. Pardo-Molina, J. Peacock, M. Peña-Claros, G.C. Pickavance, N.C.A. Pitman, L. Poorter, A. Prieto, C. A. Quesada, F. Ramírez, H. Ramírez-Angulo, Z. Restrepo, A. Roopsind, A. Rudas, R.P. Salomão, M. Schwarz, N. Silva, J.E. Silva-Espejo, M. Silveira, J. Stropp, J. Talbot, H. ter Steege, J. Teran-Aguilar, J. Terborgh, R. Thomas-Caesar, M. Toledo, M. Torello-Raventos, R.K. Umetsu, G.M.F. van der Heijden, P. van der Hout, I.C. Guimarães Vieira, S.A. Vieira, E. Vilanova, V.A. Vos and R.J. Zagt. 2015. Long term decline of the Amazon carbon sink. Nature 519: 344–349.

Bugmann, H. and C. Bigler. 2011. Will the CO_2 fertilization effect in forests be offset by reduced tree longevity? Oecologia 165: 533–544.

Bunker, D.E., F. DeClerck, J.C. Bradford, R.K. Colwell, I. Perfecto, O.L. Phillips, M. Sankaran and S. Naeem. 2005. Species loss and aboveground carbon storage in a tropical forest. Science 310: 1029–1031.

Buongiorno, J.R. Raunikar and S. Zhu. 2011. Consequences of increasing bioenergy demand on wood and forests: an application of the global forest products model. J. Forest Econ. 17: 214–229.

Canadell, J.G. and M.R. Raupach. 2008. Managing forests for climate change mitigation. Science 320: 1456–1457.

Carle, J., P. Vuorinen and A. Del Lungo. 2002. Status and trends in global forest plantation development. Forest Prod. J. 52: 1–13.

Carroll, M., B. Milakovsky, A. Finkral, A. Evans and M.S. Ashton. 2012. Managing carbon sequestration and storage in temperate and boreal forests. pp. 205–226. *In*: M.S. Ashton, M.L. Tyrrell, D. Spalding and B. Gentry (eds.). Managing Forest Carbon in a Changing Climate. Springer, Netherlands.

Catovsky, S., M.A. Bradford and A. Hector. 2002. Biodiversity and ecosystem productivity: implications for carbon storage. Oikos 97: 443–448.

Cavanaugh, K.C., J.S. Gosnell, S.L. Davis, J. Ahumada, P. Boundja, David B. Clark, B. Mugerwa, P.A. Jansen, T.G. O'Brien, F. Rovero, D. Sheil, R. Vasquez and S. Andelman. 2014. Carbon storage in tropical forests correlates with taxonomic diversity and functional dominance on a global scale. Glob. Ecol. Biogeogr. 23: 563–573.

Certini, G. 2005. Effects of fire on properties of forest soils: a review. Oecologia. 143: 1–10.

Chapin III, F.S., G.M. Woodwell, J.T. Randerson, E.B. Rastetter, G.M. Lovett, D.D. Baldocchi, D.A. Clark, M.E. Harmon, D.S. Schimel, R. Valentini, C. Wirth, J.D. Aber, J.J. Cole, M.L. Goulden, J.W. Harden, M. Heimann, R.W. Howarth, P.A. Matson, A.D. McGuire, J.M. Melillo, H.A. Mooney, J.C. Neff, R.A. Houghton, M.L. Pace, M.G. Ryan, S.W. Running, O.E. Sala, W.H. Schlesinger and E.D. Schulze. 2006. Reconciling carbon-cycle concepts, terminology and methods. Ecosystems 9: 1041–1050.

Chapin III, F.S., J. McFarland, A.D. McGuire, E.S. Euskirchen, R.W. Ruess and K. Kielland. 2009. The changing global carbon cycle: linking plant-soil carbon dynamics to global consequences. J. Ecol. 97: 840–850.

Chave, J., R. Condit, H.C. Muller-Landau, S.C. Thomas, P.S. Ashton, S. Bunyavejchewin, L.L. Co, H.S. Dattaraja, S.J. Davies, S. Esufali, C.E.N. Ewango, K.J. Feeley, R.B. Foster, N. Gunatilleke, S. Gunatilleke, P. Hall, T.B. Hart, C. Hernández, S.P. Hubbell, A. Itoh, S. Kiratiprayoon, J.V. LaFrankie, S. Loo de Lao, J.R. Makana, Md. Nur Supardi Noor, A. Rahman Kassim, C. Samper, R. Sukumar, H.S. Suresh, S. Tan, J. Thompson, Ma. D.C. Tongco, R. Valencia, M. Vallejo, G. Villa, T. Yamakura, J.K. Zimmerman and E.C. Losos. 2008. Assessing evidence for a pervasive alteration in tropical tree communities. PLoS Biology 6: e45.

Chen, G., Y. Yang and D. Robinson. 2014. Allometric constraints on and trade-offs in belowground carbon allocation and their control of soil respiration across global forest ecosystems. Glob. Change Biol. 20: 1674–1684.

Churkina, G., K. Trusilova, M. Vetter and F.J. Dentener. 2007. Contributions of nitrogen deposition and forest re-growth to land carbon uptake. Carbon Balance and Management 2: 5.

Clark, D.A. 2004. Tropical forests and global warming: slowing it down or speeding it up? Front. Ecol. Environ. 2: 73–80.

Clark, D.A., S. Brown, D.W. Kicklighter, J.Q. Chambers, J.R. Thomlinson, J. Ni and E.A. Holland. 2001. Net primary production in tropical forests: an evaluation and synthesis of existing field data. Ecol. Appl. 11: 317–384.

Clark, D.B., D.A. Clark, S. Brown, S.F. Oberbauer and E. Veldkamp. 2002. Stocks and flows of coarse woody debris across a tropical rainforest nutrient and topography gradient. Forest Ecol. Manag. 164: 237–248.

Clark, D.B., D.A. Clark and S.F. Oberbauer. 2010. Annual wood production in a tropical rain forest in NE Costa Rica linked to climatic variation but not to increasing CO_2. Glob. Change Biol. 16: 747–759.

Conant, R.T., J.M. Steinweg, M.L. Haddix, E.A. Paul, A.F. Plante and J. Six. 2008. Experimental warming shows that decomposition temperature sensitivity increases with soil carbon recalcitrance. Ecology 89: 2384–2391.

Cox, P.M., R.A. Betts, M. Collins, P.P. Harris, C. Huntingford and C.D. Jones. 2004. Amazonian forest dieback under climate-carbon cycle projections for the 21st century. Theor. Appl. Climatol. 78: 137–156.

Cramer, W., A. Bondeau, F.I. Woodward, C. Prentice, R.A. Betts, V. Brovkin, P.M. Cox, V. Fisher, J.A. Foley, A.D. Friend, C. Kucharik, M.R. Lomas, N. Ramankutty, S. Sitch, B. Smith, A. White and C. Young-Molling. 2001. Global response of terrestrial ecosystem structure and function to CO_2 and climate change: results from six dynamic global vegetation models. Glob. Change Biol. 7: 357–374.

Czimczik, C.I. and C.A. Masiello. 2007. Controls on black carbon storage in soils. Global Biogeochem. Cy. 21: B3005.

Czimczik, C.A., S.E. Trumbore, M.S. Carbone and G.C. Winston. 2006. Changing sources of soil respiration with time since fire in a boreal forest. Glob. Change Biol. 12: 957–971.

De Deyn, G.B., J.H.C. Cornelissen and R.D. Bardgett. 2008. Plant functional traits and soil carbon sequestration in contrasting biomes. Ecol. Lett. 11: 516–531.

De Vries, W., G.J. Reinds, P. Gundersen and H. Sterba. 2006. The impact of nitrogen deposition on carbon sequestration in European forests and forest soils. Glob. Change Biol. 12: 1151–1173.

Dean, C., S.H. Roxburgh, R.J. Harper, D.J. Eldridge, I.W. Watson and G.W. Wardell-Johnson. 2012. Accounting for space and time in soil carbon dynamics in timbered rangelands. Ecol. Eng. 38: 51–64.

DeLuca, T.H. and G.H. Aplet. 2008. Charcoal and carbon storage in forest soils of the Rocky Mountain West. Front. Ecol. Environ. 6: 18–24.

DeLuca, T.H. and C. Boisvenue. 2012. Boreal forest soil carbon: distribution, function and modelling. Forestry 85: 161–184.

Del Cid-Liccardi, C., T. Kramer, M.S. Ashton and B. Griscom. 2012. Managing carbon sequestration in tropical forests. pp. 183–204. *In*: M.S. Ashton, M.L. Tyrrell, D. Spalding and B. Gentry (eds.). Managing Forest Carbon in a Changing Climate. Springer, Netherlands.

Denman, K.L., G. Brasseur, A. Chidthaisong, P. Ciais, P.M. Cox, R.E. Dickinson, D. Hauglustaine, C. Heinze, E. Holland, D. Jacob, U. Lohmann, S. Ramachandran, P.L. da Silva Dias, S.C. Wofsy and X. Zhang. 2007. Couplings between changes in the climate system and biogeochemistry. pp. 499–587. *In*: S. Solomon, D. Qin, M. Manning, Z. Chen, M. Marquis, K.B. Averyt, M. Tignor and H.L. Miller (eds.). Climate Change 2007: The Physical Science Basis. Contribution of Working Group I to the Fourth Assessment Report of the Intergovernmental Panel on Climate Change. Cambridge University Press, Cambridge, United Kingdom.

Don, A., J. Schumacher and A. Freibauer. 2011. Impact of tropical land-use change on soil organic carbon stocks—a meta-analysis. Glob. Change Biol. 17: 1658–1670.

Donato, D.C., J.B. Kauffman, D. Murdiyarso, S. Kurnianto, M. Stidham and M. Kanninen. 2011. Mangroves among the most carbon rich forests in the tropics. Nat. Geosci. 4: 293–297.

Dutaur, L. and L.V. Verchot. 2007. A global inventory of the soil CH_4 sink. Global Biogeochem. Cy. 21: GB4013.

Edburg, S.L., J.A. Hicke, P.D. Brooks, E. Pendall, B.E. Ewers, U. Norton, D. Gochis, E.D. Gutmann and A.J.H. Meddens. 2012. Cascading impacts of bark beetle-caused tree mortality on coupled biogeophysical and biogeochemical processes. Front. Ecol. Environ. 10: 416–424.

Erskine, P.D., D. Lamb and M. Bristow. 2006. Tree species diversity and ecosystem function: can tropical multispecies plantations generate greater productivity? Forest Ecol. Manag. 233: 205–210.

Fahey, T.J., P.B. Woodbury, J.J. Battles, C.L. Goodale, S.P. Hamburg, S.V. Ollinger and C.W. Woodall. 2010. Forest carbon storage: ecology, management, and policy. Front. Ecol. Environ. 8: 245–252.

[FAO] Food and Agriculture Organization of the United Nations. 1993. Forest Resources Assessment— Tropical Countries. FAO Forestry Paper 112. Rome.

[FAO] Food and Agriculture Organization of the United Nations. 2010. Global Forest Resources Assessment. FAO Forestry Paper 163. Rome.

[FAO] Food and Agriculture Organization of the United Nations. 2011. State of the World's Forests. Rome.

[FAO] Food and Agriculture Organization of the United Nations. 2012. State of the World's Forests. Rome.

Friedlingstein, P., R.A. Houghton, G. Marland, J. Hackler, T.A. Boden, T.J. Conway, J.G. Canadell, M.R. Raupach, P. Ciais and C. Le Quéré. 2010. Update on CO_2 emissions. Nat. Geosci. 3: 811–812.

Galloway, J.N., A.R. Townsend, J.W. Erisman, M. Bekunda, Z. Cai, J.R. Freney, L.A. Martinelli, S.P. Seitzinger and M.A. Sutton. 2008. Transformation of the nitrogen cycle: recent trends, questions, and potential solutions. Science 320: 889–892.

Global Carbon Project. 2014. Global carbon budget 2014. www.globalcarbonproject.org

Gorham, E. 1991. Northern peatlands: role in the carbon cycle and probable responses to climate warming. Ecol. Appl. 1: 182–195.

Goulden, M.L., S. Wofsy, J.W. Harden, S.E. Trumbore, P.M. Crill, S.T. Gower, T. Fries, B.C. Daube, S.M. Fan, D.J. Sutton, A. Bazzaz and J.W. Munger. 1998. Sensitivity of boreal forest carbon balance to soil thaw. Science 279: 214–217.

Goulden, M.L., A.M.S. McMillan, G.C. Winston, A.V. Rocha, K.L. Manies, J.W. Harden and B.P. Bond-Lamberty. 2011. Patterns of NPP, GPP, respiration, and NEP during boreal forest succession. Glob. Change Biol. 17: 855–871.

Gower, S.T., O. Krankina, R.J. Olson, M. Apps, S. Linder and C. Wang. 2001. Net primary production and carbon allocation patterns of boreal forest ecosystems. Ecol. Appl. 11: 1395–1411.

Gower, S.T., J.J. Landsberg and K.E. Bisbee. 2003. Forest biomes of the world. pp. 57–74. *In*: R.A. Young and R.L. Giese (eds.). Forest Ecosystem Science and Management. Wiley, Hoboken, USA.

Grace, J. 2004. Understanding and managing the global carbon cycle. J. Ecol. 92: 189–202.

Grace, J., J. San José, P. Meir, H.S. Miranda and R.A. Montes. 2006. Productivity and carbon fluxes of tropical savannas. J. Biogeogr. 33: 387–400.

Griscom, B., P. Ellis and F.E. Putz. 2014. Carbon emission performance of commercial logging in East Kalimantan, Indonesia. Glob. Change Biol. 20: 923–937.

Guenther, A., T. Karl, P. Harley, C. Wiedinmeyer, P.I. Palmer and C. Geron. 2006. Estimates of global terrestrial isoprene emissions using MEGAN (model of emissions of gases and aerosols from nature). Atmos. Chem. Phys. 6: 3181–3210.

Guggenberger, G., A. Rodionov, O. Shibistova, M. Grabe, O.A. Kasansky and H. Fuchs. 2008. Storage and mobility of black carbon in permafrost soils of the forest tundra ecotone in northern Siberia. Glob. Change Biol. 14: 1367–1381.

Hansen, M.C., S.V. Stehman, P.V. Potapov, T.R. Loveland, J.R.G. Townshend, R.S. De Fries, K.W. Pittman, B. Arunarwati, F. Stolle, M.K. Steininger, M. Carroll and C. Di Miceli. 2008. Humid tropical forest clearing from 2000 to 2005 quantified by using multitemporal and multiresolution remotely sensed data. P. Natl. Acad. Sci. USA 105: 9439–9444.

Hansen, M.C., P.V. Potapov, R. Moore, M. Hancher, S.A. Turubanova, A. Tyukavina, D. Thau, S.V. Stehman, S.J. Goetz, T.R. Loveland, A. Kommareddy, A. Egorov, L. Chini, C.O. Justice and J.R.G. Townshend. 2013. High-resolution global maps of 21st-century forest cover change. Science 343: 850–853.

Harris, N.L., S. Brown, S.C. Hagen, S.S. Saatchi, S. Petrova, W. Salas, M.C. Hansen, P.V. Potapov and A. Lotsch. 2012. Baseline map of carbon emissions from deforestation in tropical regions. Science 336: 1573–1576.

Harvey, C.A., M. Chacón, C.I. Donatti, E. Garen, L. Hannah, A. Andrade, L. Bede, D. Brown, A. Calle, J. Chará, C. Clement, E. Gray, M.H. Hoang, P. Minang, A.M. Rodríguez, C. Seeberg-Elverfeldt, B. Semroc, S. Shames, S. Smukler, E. Somarriba, E. Torquebiau, J. van Etten and E. Wollenberg. 2014. Climate-smart landscapes: opportunities and challenges for integrating adaptation and mitigation in tropical agriculture. Conserv. Lett. 7: 77–90.

Hayes, D.J., A.D. McGuire, D.W. Kicklighter, K.R. Gurney, T.J. Burnside and J.M. Melillo. 2011. Is the northern high-latitude land-based CO_2 sink weakening? Global Biogeochem. Cy. 25: GB3018.

Heimann, M. and M. Reichstein. 2008. Terrestrial ecosystem carbon dynamics and climate feedbacks. Nature 451: 289–292.

Högberg, P. 1997. Tansley review no. 95-^{15}N natural abundance in soil-plant systems. New Phytol. 137: 179–203.

Houghton, R.A. 1991. Releases of carbon to the atmosphere from degradation of forests in tropical Asia. Can. J. Forest Res. 21: 132–142.

Houghton, R.A., J.I. House, J. Pongratz, G.R. van der Werf, R.S. DeFries, M.C. Hansen, C. Le Quéré and N. Ramankutty. 2012. Carbon emissions form land use and land-cover change. Biogeosciences 9: 5125–5142.

Huntingford, C., P. Zelazowski, D. Galbraith, L.M. Mercado, S. Sitch, R. Fisher, M. Lomas, A.P. Walker, C.D. Jones, B.B.B. Booth, Y. Malhi, D. Hemming, G. Kay, P. Good, S.L. Lewis, O.L. Phillips, O.K. Atkin, J. Lloyd, E. Gloor, J. Zaragoza-Castells, P. Meir, R. Betts, P.P. Harris, C. Nobre, J. Marengo and P.M. Cox. 2013. Simulated resilience of tropical rainforests to CO_2-induced climate change. Nat. Geosci. 6: 268–273.

Hyvönen, R., G.I. Ågren, S. Linder, T. Persson, M.F. Cotrufo, A. Ekblad, M. Freeman, A. Grelle, I.A. Janssens, P.G. Jarvis, S. Kellomäki, A. Lindroth, D. Loustau, T. Lundmark, R.J. Norby, R. Oren, K. Pilegaard, M.G. Ryan, B.D. Sigurdsson, M. Strömgren, M. van Oijen and G. Wallin. 2007. The likely impact of elevated [CO_2], nitrogen deposition, increased temperature and management on carbon sequestration in temperate and boreal forest ecosystems: a literature review. New Phytol. 173: 463–480.

Inubushi, K., Y. Furukawa, A. Hadi, E. Purnomo and H. Tsuruta. 2003. Seasonal changes of CO_2, CH_4, and N_2O fluxes in relation to land-use change in tropical peatlands located in coastal area of South Kalimantan. Chemosphere 52: 603–608.

[IPCC] Intergovernmental Panel on Climate Change. 2014. Climate Change 2014: Synthesis Report. Contribution of Working Groups I, II and III to the Fifth Assessment Report of the Intergovernmental Panel on Climate Change [Core Writing Team, R.K. Pachauri and L.A. Meyer (eds.)]. Geneva, Switzerland, 151pp.

Jandl, R., M. Lindner, L. Vesterdal, B. Bauwens, R. Baritz, F. Hagedorn, D.W. Johnson, K. Minkkinen and K.A. Byrne. 2007. How strongly can forest management influence soil carbon sequestration? Geoderma 137: 253–268.

Jobbagy, E.G. and R.B. Jackson. 2000. The vertical distribution of soil organic carbon and its relation to climate and vegetation. Ecol. Appl. 10: 423–436.

Johnson, K., F.N. Scatena and Y. Pan. 2010. Short- and long-term responses of total soil organic carbon to harvesting in a northern hardwood forest. Forest Ecol. Manag. 259: 1262–1267.

Jose, S. and S. Bardhan. 2012. Agroforestry for biomass production and carbon sequestration: an overview. Agroforest. Syst. 86: 105–111.

Kapos, V., W.A. Kurz, T. Gardner, J. Ferreira, M. Guariguata, L.P. Koh, S. Mansourian, J.A. Parrotta, N. Sasaki, C.B. Schmitt, J. Barlow, M. Kanninen, K. Okabe, Y. Pan, I.D. Thompson and N. van Vliet. 2012. Impacts of forest and land management on biodiversity and carbon. pp. 53–80. *In*: J.A. Parrotta, C. Wildburger and S. Mansourian (eds.). Understanding Relationships between Biodiversity, Carbon, Forests and People: The Key to Achieving REDD+ Objectives. International Union of Forest Research Organizations, Vienna, Austria.

Kasischke, E.S. and B.J. Stocks. 2000. Fire, Climate Change, and Carbon Cycling in the Boreal Forest. Ecological Studies 138. Springer-Verlag, New York, USA.

Kaspari, M., M.N. Garcia, K.E. Harms, M. Santana, S.J. Wright and J.B. Yavitt. 2008. Multiple nutrients limit litterfall and decomposition in a tropical forest. Ecol. Lett. 11: 35–43.

Keith, H., B.G. Mackey and D.B. Lindenmayer. 2009. Re-evaluation of forest biomass carbon stocks and lessons from the world's most carbon-dense forests. P. Natl. Acad. Sci. 106: 11635–11640.

Kim, D.H., J.O. Sexton and J.R. Townshend. 2015. Accelerated deforestation in the humid tropics from the 1990s to the 2000s. Geophys. Res. Lett. 41, doi: 10.1002/2014GL062777.

Keppler, F., J.T.G. Hamilton, M. Braß and T. Röckmann. 2006. Methane emissions from terrestrial plants under aerobic conditions. Nature 439: 187–191.

Kirby, K.R. and C. Potvin. 2007. Variation in carbon storage among tree species: implications for the management of a small scale carbon sink project. Forest Ecol. Manag. 246: 208–221.

Kormos, C.F. and B.L. Zimmerman. 2014. Response to: Putz et al., Sustaining conservation values in selectively logged tropical forests: the attained and the attainable. Conservation Letters 7: 143–144.

Kraenzel, M., A. Castillo, T. Moore and C. Potvin. 2003. Carbon storage of harvest-age teak (*Tectona grandis*) plantations, Panama. Forest Ecol. Manag. 173: 213–225.

Kumar, B.M. and P.K.R. Nair (eds.). 2011. Carbon Sequestration Potential of Agroforestry Systems—Opportunities and Challenges. Springer, Netherlands.

Kurz, W.A., G. Stinson, G.J. Rampley, C.C. Dymond and E.T. Neilson. 2008. Risk of natural disturbances makes future contribution of Canada's forest to the global carbon cycle highly uncertain. P. Natl. Acad. Sci. USA 105: 1551–1555.

Larjavaara, M. 2008. A review of benefits and disadvantages of tree diversity. The Open Forest Science Journal 1: 24–26.

Larsen, J.A. 1980. The Boreal Forest Ecosystem. Academic Press, New York, USA.

Laurance, W.F., A.A. Oliveira, S.G. Laurance, R. Condit, H.E.M. Nascimento, A.C. Sanchez-Thorin, T.E. Lovejoy, A. Andrade, S. D'Angelo, J.E. Ribeiro and C.W. Dick. 2004. Pervasive alteration of tree communities in undisturbed Amazonian forests. Nature 428: 171–175.

Laurance, W.F., A.S. Andrade, A. Magrach, J.L.C. Camargo, J.J. Valsko, M. Campbell, P.M. Fearnside, W. Edwards, T.E. Lovejoy and S.G. Laurance. 2014. Long-term changes in liana abundance and forest dynamics in undisturbed Amazonian forests. Ecology 95: 1604–1611.

Le Quéré, C., M.R. Raupach, J.G. Canadell, G. Marland, L. Bopp, P. Ciais, T.J. Conway, S.C. Doney, R.A. Feely, P. Foster, P. Friedlingstein, K. Gurney, R.A. Houghton, J.I. House, C. Huntingford, P.E. Levy, M.R. Lomas, J. Majkut, N. Metzl, J.P. Ometto, G.P. Peters, I.C. Prentice, J.T. Randerson, S.W. Running, J.L. Sarmiento, U. Schuster, S. Sitch, T. Takahashi, N. Viovy, G.R. van der Werf and F.I. Woodward. 2009. Trends in the sources and sinks of carbon dioxide. Nat. Geosci. 2: 831–836.

Le Quéré, C., R. Moriarty, R.M. Andrew, G.P. Peters, P. Ciais, P. Friedlingstein, S.D. Jones, S. Sitch, P. Tans, A. Arneth, T.A. Boden, L. Bopp, Y. Bozec, J.G. Canadell, F. Chevallier, C.E. Cosca, I. Harris, M. Hoppema, R.A. Houghton, J.I. House, A. Jain, T. Johannessen, E. Kato, R.F. Keeling, V. Kitidis, K. Klein Goldewijk, C. Koven, C.S. Landa, P. Landschützer, A. Lenton, I.D. Lima, G. Marland, J.T. Mathis, N. Metzl, Y. Nojiri, A. Olsen, T. Ono, W. Peters, B. Pfeil, B. Poulter, M.R. Raupach, P. Regnier, C. Rödenbeck, S. Saito, J.E. Salisbury, U. Schuster, J. Schwinger, R. Séférian, J. Segschneider, T. Steinhoff, B.D. Stocker, A.J. Sutton, T. Takahashi, B. Tilbrook, G.R. van der Werf, N. Viovy, Y.P. Wang, R. Wanninkhof, A. Wiltshire and N. Zeng. 2014. Global carbon budget. Earth Syst. Sci. Data Discuss 7: 521–610.

Lewis, S.L., G. Lopez-Gonzalez, B. Sonké, K. Affum-Baffoe, T.R. Baker, L.O. Ojo, O.L. Phillips, J.M. Reitsma, L. White, J.A. Comiskey, M.N. Djuikouo K., C.E.N. Ewango, T.R. Feldpausch, A.C. Hamilton, M. Gloor, T. Hart, A. Hladik, J. Lloyd, J.C. Lovett, J.R. Makana, Y. Malhi, F.M. Mbago, H.J. Ndangalasi, J. Peacock, K.S.H. Peh, D. Sheil, T. Sunderland, M.D. Swaine, J. Taplin, D. Taylor, S.C. Thomas, R. Votere and H. Wöll. 2009. Increasing carbon storage in African tropical forests. Nature 457: 1003–1007.

Liao, C., Y. Luo, C. Fang and B. Li. 2010. Ecosystem carbon stock influenced by plantation practice: implications for planting forests as a measure of climate change mitigation. PLoS ONE 5: e10867.

Lichstein, J.W., C. Wirth, H.S. Horn and S.W. Pacala. 2009. Biomass chronosequences of United States forests: implications for carbon storage and forest management. pp. 301–341. *In*: C. Wirth, G. Gleixner and M. Heimann (eds.). Old-Growth Forests: Function, Fate and Value. Springer, Berlin Heidelberg, Germany.

Litton, C.M., J.W. Raich and M.G. Ryan. 2007. Carbon allocation in forest ecosystems. Glob. Change Biol. 13: 2089–2109.

Logan-Hines, E., L. Goers, M. Evidente and B. Cashore. 2012. REDD+ policy options: including forests in an international climate change agreement. pp. 357–376. *In*: M.S. Ashton, M.L. Tyrrell, D. Spalding and B. Gentry (eds.). Managing Forest Carbon in a Changing Climate. Springer, Netherlands.

Lorenz, K. and R. Lal. 2010. Carbon Sequestration in Forest Ecosystems. Springer, Netherlands.

Luyssaert, S., I. Inglima, M. Jung, A.D. Richardson, M. Reichstein, D. Papale, S.L. Piao, E.D. Schulze, L. Wingate, G. Matteucci, L. Aragao, M. Aubinet, C. Beer, C. Bernhofer, K.G. Black, D. Bonan, J.M. Bonnefond, J. Chambers, P. Ciais, B. Cook, K.J. Davis, A.J. Dolman, B. Gielen, M. Goulden, J. Grace, A. Granier, A. Grelle, T. Griffis, T. Grunwald, G. Guidolotti, P.J. Hanson, R. Harding, D.Y. Hollinger, L.R. Hutyra, P. Kolar, B. Kruijt, W. Kutsch, F. Lagergren, T. Laurila, B.E. Law, G. Le Maire, A. Lindroth, D. Loustau, Y. Mahli, J. Mateus, M. Migliavacca, L. Mission, L. Montagnani, J. Moncrieff, E. Moors, J.W. Munger, E. Nikinmaa, S.V. Ollinger, G. Pita, C. Rebmann, O. Roupsard, N. Saigusa, M.J. Sanz, G. Seufert, C. Sierra, M.-L. Smith, J. Tang, R. Valentini, T. Vesala and I.A. Janssens. 2007. CO_2 balance of boreal, temperate, and tropical forests derived from a global database. Glob. Change Biol. 13: 2509–2537.

Luyssaert, S., E.D. Schulze, A. Börner, A. Knohl, D. Hessenmöller, B.E. Law, P. Ciais and J. Grace. 2008. Old-growth forests as carbon sinks. Nature 455: 213–215.

Magnani, F., M. Mencuccini, M. Borghetti, P. Berbigier, F. Berninger, S. Delzon, A. Grelle, P. Hari, P.G. Jarvis, P. Kolari, A.S. Kowalski, H. Lankreijer, B.E. Law, A. Lindroth, D. Loustau, G. Manca, J.B. Moncrieff, M. Rayment, V. Tedeschi, R. Valentini and J. Grace. 2007. The human footprint in the carbon cycle of temperate and boreal forests. Nature 447: 848–850.

Mahli, Y. 2010. The carbon balance of tropical forest regions, 1990–2005. Curr. Opin. Env. Sust. 2: 237–244.

Mahli, Y. 2012. The productivity, metabolism and carbon cycle of tropical forest vegetation. J. Ecol. 100: 65–75.

Mahli, Y. and J. Grace. 2000. Tropical forests and atmospheric carbon dioxide. Trends Ecol. Evol. 15: 332–337.

Mahli, Y., C. Doughty and D. Galbraith. 2011. The allocation of ecosystem net primary productivity in tropical forests. Philos. Trans. Roy. Soc. B 366: 3225–3245.

Mascarelli, A.L. 2009. A sleeping giant? Nat. Rep. Clim. Change 3: 46–49.

Mazzei, L., P. Sist, A. Ruschel, F.E. Putz, P. Marco, W. Pena and J.E. Ribeiro Ferreira. 2010. Above-ground biomass dynamics after reduced-impact logging in the Eastern Amazon. Forest Ecol. Manag. 259: 367–373.

McDowell, N.G., W. Pockman, C. Allen, D. Breshears, N. Cobb, T. Kolb, J. Plaut, J. Sperry, A. West, D. Williams and E. Yepez. 2008. Mechanisms of plant survival and mortality during drought: why do some plants survive while others succumb? New Phytol. 178: 719–739.

[MCPFE] Ministerial Conference on the Protection of Forests in Europe. 2007. State of Europe's forests 2007. The MCPFE report on Sustainable Forest Management in Europe. MCPFE Liaison Unit Warsaw, United Nations Economic Commission for Europe and Food and Agriculture Organization of the United Nations. Warsaw, Poland, 247pp.

Meister, K., M.S. Ashton, D. Craven and H. Griscom. 2012. Carbon dynamics of tropical forests. pp. 51–75. *In*: M.S. Ashton, M.L. Tyrrell, D. Spalding and B. Gentry (eds.). Managing Forest Carbon in a Changing Climate Springer, Netherlands.

Midgley, G.F., W.J. Bond, V. Kapos, C. Ravilious, J.P.W. Scharlemann and F.I. Woodward. 2010. Terrestrial carbon stocks and biodiversity: key knowledge gaps and some policy implications. Curr. Opin. Env. Sust. 2: 264–270.

Mobley, M.L. 2011. An ecosystem approach to dead plant carbon over 50 years of old-field forest development. Ph.D. Thesis, Duke University, Durham, North Carolina.

Moss, R.H., J.A. Edmonds1, K.A. Hibbard, M.R. Manning, S.K. Rose, D.P. van Vuuren, T.R. Carter, S. Emori, M. Kainuma, T. Kram, G.A. Meehl, J.F.B. Mitchell, N. Nakicenovic, K. Riahi, S.J. Smith, R.J. Stouffer, A.M. Thomson, J.P. Weyant and T.J. Wilbanks. 2010. The next generation of scenarios for climate change research and assessment. Nature 463: 747–756.

Müller, M.J. 1982. Selected Climate Data for a Global Set of Standard Set of Stations for Vegetation Science. Springer, Netherlands.

Murphy, P.G. and A.E. Lugo. 1986. Ecology of tropical dry forest. Annu. Rev. Ecol. Syst. 17: 67–88.

Nabuurs, G.J., O. Masera, K. Andrasko, P. Benitez-Ponce, R. Boer, M. Dutschke, E. Elsiddig, J. Ford-Robertson, P. Frumhoff, T. Karjalainen, O. Krankina, W.A. Kurz, M. Matsumoto, W. Oyhantcabal, N.H. Ravindranath, M.J. Sanz Sanchez and X. Zhang. 2007. Forestry. pp. 541–584. *In*: B. Metz, O.R. Davidson, P.R. Bosch, R. Dave and L.A. Meyer (eds.). Climate Change 2007: Mitigation. Contribution of Working Group III to the Fourth Assessment Report of the Intergovernmental Panel on Climate Change. Cambridge University Press, Cambridge, United Kingdom.

Nabuurs, G.J., M. Lindner, P.J. Verkerk, K. Gunia, P. Deda, R. Michalak and G. Grassi. 2013. First signs of carbon sink saturation in European forest biomass. Nat. Clim. Change 3: 792–796.

Nieder, R. and D.K. Benbi. 2008. Carbon and Nitrogen in the Terrestrial Environment. Springer, Berlin, Germany.

Norby, R.J., J.M. Warren, C.M. Iversen, B.E. Medlyn and R.E. McMurtrie. 2010. CO_2 enhancement of forest productivity constrained by limited nitrogen availability. P. Natl. Acad. Sci. USA 107: 19368–19373.

Odum, E.P. 1969. The strategy of ecosystem development. Science 164: 262–270.

Pan, Y., R.A. Birdsey, J. Fang, R.A. Houghton, P.E. Kauppi, W.A. Kurz, O.L. Phillips, A. Shvidenko, S.L. Lewis, J.G. Canadell, P. Ciais, R.B. Jackson, S.W. Pacala, A.D. McGuire, S. Piao, A. Rautiainen, S. Sitch and D. Hayes. 2011. A large and persistent carbon sink in the world's forests. Science 333: 988–993.

Pan, Y., R.A. Birdsey, O.L. Phillips and R.B. Jackson. 2013. The structure, distribution and biomass of the world's forests. Annu. Rev. Ecol. Syst. 44: 593–622.

Peng, C.H. and J.M. Apps. 1999. Modeling the response of net primary productivity (NPP) of Boreal Forest ecosystems to changes in climate and fire disturbance regimes. Ecol. Model. 122: 175–193.

Penman, J., M. Gytarsky, T. Hiraishi, T. Krug, D. Kruger, R. Pipatti, L. Buendia, K. Miwa, T. Ngara, K. Tanabe and F. Wagner. 2003. Intergovernmental Panel on Climate Change. Good Practice Guidance for Land Use, Land-Use Change and Forestry. Institute for Global Environmental Strategies, Kanagawa, Japan.

Phillips, O.L., S.L. Lewis, T.R. Baker, K.J. Chao and N. Higuchi. 2008. The changing Amazon forest. Philos. Trans. R. Soc. B 363: 1819–1828.

Piao, S., S. Sitch, P. Ciais, P. Friedlingstein, P. Peylin, X. Wang, A. Ahlström, A. Anav, J.G. Canadell, N. Cong, C. Huntingford, M. Jung, S. Levis, P.E. Levy, J. Li, X. Lin, M.R. Lomas, M. Lu, Y. Luo, Y. Ma, R.B. Myneni, B. Poulter, Z. Sun, T. Wang, N. Viovy, S. Zaehle and N. Zeng. 2013. Evaluation of terrestrial carbon cycle models for their response to climate variability and CO_2 trends. Glob. Change Biol. 19: 2117–2132.

Ping, C.L., G.J. Michaelson, E.S. Kane, E.C. Packee, C.A. Stiles, D.K. Swanson and N.D. Zaman. 2010. Carbon stores and biogeochemical properties of soils under black spruce forest, Alaska. Soil Sci. Soc. Am. J. 74: 969–978.

Potvin, C., L. Mancilla, N. Buchmann, J. Monteza, T. Moore, M. Murphy, Y. Oelmann, M. Scherer-Lorenzen, B.L. Turner, W. Wilcke, F. Zeugin and S. Wolf. 2011. An ecosystem approach to biodiversity effects: carbon pools in a tropical tree plantation. Forest Ecol. Manag. 261: 1614–1624.

Pregitzer, K.S. and E.S. Euskirchen. 2004. Carbon cycling and storage in world forests: biome patterns related to forest age. Glob. Change Biol. 10: 2052–2077.

Prentice, I.C., G.D. Farquhar, M.J.R. Fasham, M.L. Goulden, M. Heimann, V.J. Jaramillo, H.S. Kheshgi, C. Le Quéré, R.J. Scholes, D.W.R. Wallace, D. Archer, M.R. Ashmore, O. Aumont, D. Baker, M. Battle, M. Bender, L.P. Bopp, P. Bousquet, K. Caldeira, P. Ciais, P.M. Cox, W. Cramer, F. Dentener, I.G. Enting, C.B. Field, P. Friedlingstein, E.A. Holland, R.A. Houghton, J.I. House, A. Ishida, A.K. Jain, I.A. Janssens, F. Joos, T. Kaminski, C.D. Keeling, R.F. Keeling, D.W. Kicklighter, K.E. Kohfeld, W. Knorr, R. Law, T. Lenton, K. Lindsay, E. Maier-Reimer, A.C. Manning, R.J. Matear, A.D. McGuire, J.M. Melillo, R. Meyer, M. Mund, J.C. Orr, S. Piper, K. Plattner, P.J. Rayner, S. Sitch, R. Slater, S. Taguchi, P.P. Tans, H.Q. Tian, M.F. Weirig, T. Whorf and A. Yool. 2001. The carbon cycle and atmospheric carbon dioxide. pp. 183–227. *In*: J.T. Houghton, Y. Ding, D.J. Griggs, M. Noguer, P.J. van der Linden, X. Dai, K. Maskell and C.A. Johnson (eds.). Climate Change 2001: The Scientific Basis. Contribution of Working Group I to the Third Assessment Report of the Intergovernmental Panel on Climate Change. Cambridge University Press, Cambridge, United Kingdom.

Preston, C.M. and M.W.I. Schmidt. 2006. Black (pyrogenic) carbon: a synthesis of current knowledge and uncertainties with special consideration of boreal regions. Biogeoscience 3: 397–420.

Pretzsch, H. and G. Schütze. 2009. Transgressive over yielding in mixed compared with pure stands of Norway spruce and European beech in Central Europe: evidence on stand level and explanation on individual tree level. Eur. J. Forest Res. 128: 183–204.

Profft, I., M. Mund, G.E. Weber, E. Weller and E.D. Schulze. 2009. Forest management and carbon sequestration in wood products. Eur. J. Forest Res. 128: 399–413.

Putz, F.E. 2013. Complexity confronting tropical silviculturists. pp. 165–186. *In*: C. Messier, J. Puettman and K.D. Coates (eds.). Managing Forests as Complex Adaptive Systems: Building Resilience to the Challenge of Global Change. Earthscan, New York, USA.

Putz, F.E., P.A. Zuidema, M.A. Pinard, R.G.A. Boot, J.A. Sayer, D. Sheil, P. Sist, Elias and J.K. Vanclay. 2008a. Improved tropical forest management for carbon retention. PLoS Biol. 6: e166.

Putz, F.E., P. Sist, T. Fredericksen and D. Dykstra. 2008b. Reduced impact logging: challenges and opportunities. Forest Ecol. Manag. 256: 1427–1433.

Reich, P.B. 2011. Taking stock of forest carbon. Nat. Clim. Change 1: 346–347.

Reijnders, L. and M.A.J. Huijbregts. 2008. Palm oil and the emission of greenhouse gases. J. Clean. Prod. 16: 477–482.

Resco de Dios, V., C. Colinas and C. Fischer. 2007. Climate change effects on Mediterranean forests and preventive measures. New Forests 33: 29–40.

Resh, S.C., D. Binkley and J.A. Parrotta. 2002. Greater soil carbon sequestration under nitrogen-fixing trees compared with *Eucalyptus* species. Ecosystems 5: 217–231.

Robinson, D. 2007. Implications of a large global root biomass for carbon sink estimates. Proc. R. Soc. B 274: 2753–2759.

Rockström, J., W. Steffen, K. Noone, Å. Persson, F.S. Chapin III, E.F. Lambin, T.M. Lenton, M. Scheffer, C. Folke, H.J. Schellnhuber, B. Nykvist, C.A. de Wit, T. Hughes, S. van der Leeuw, H. Rodhe, S. Sörlin, P.K. Snyder, R. Costanza, U. Svedin, M. Falkenmark, L. Karlberg, R.W. Corell, V.J. Fabry, J. Hansen, B. Walker, D. Liverman, K. Richardson, P. Crutzen and J.A. Foley. 2009. A safe operating space for humanity. Nature 461: 472–475.

Ruiz-Jaen, M.C. and C. Potvin. 2010. Can we predict carbon stocks in tropical ecosystems from tree diversity? Comparing species and functional diversity in a plantation and a natural forest. New Phytol. 189: 978–987.

Sabine, C.L., M. Heimann, P. Artaxo, D.C.E. Bakker, C.T.A. Chen, C.B. Field, N. Gruber, C. Le Quéré, R.G. Prinn, J.E. Richey, P. Romero Lankao, J.A. Sathaye and R. Valentini. 2004. Current and past trends of the global carbon cycle. pp. 17–44. *In*: C.B. Field and M.R. Raupach (eds.). The Global Carbon Cycle: Integrating Humans, Climate, and the Natural World. Island Press, Washington D.C., USA.

Sanchez, F.G., M. Coleman, C.T. Garten, Jr., R.J. Luxmoore, J.A. Stanturf, C. Trettin and S.D. Wullschleger. 2007. Soil carbon, after 3 years, under short-rotation woody crops grown under varying nutrient and water availability. Biomass Bioenerg. 31: 793–801.

Sang, P.M., D. Lamb, M. Bonner and S. Schmidt. 2013. Carbon sequestration and soil fertility of tropical tree plantations and secondary forest established on degraded land. Plant Soil 362: 187–200.

Saugier, B., J. Roy and H.A. Mooney. 2001. Estimates of global terrestrial productivity: converging toward a single number? pp. 543–557. *In*: J. Roy, B. Saugier and H.A. Mooney (eds.). Terrestrial Global Productivity. Academic Press, San Diego, USA.

Schulze, E.D., J. Lloyd, F.M. Kelliher, C. Wirth, C. Rebmann, B.M. Lühker, M. Mund, A. Knoh, I.M. Milyukova, W. Schulze, W. Ziegler, A.V. Varlagin, A.F. Sogachev, R. Valentini, S. Dore, S. Grigoriev, O. Kolle, M.I. Panfyorov, N.M. Tchebakova and N.N. Vygodskaya. 1999. Productivity of forests in the Eurosiberian boreal region and their potential to act as a carbon sink—a synthesis. Glob. Change Biol. 5: 703–722.

Schulze, E.D., D. Hessenmoeller, A. Knohl, S. Luyssaert, A. Boerner and J. Grace. 2009. Temperate and boreal old-growth forests: how do their growth dynamics and biodiversity differ from young stands and managed forests? pp. 343–366. *In*: C. Wirth, G. Gleixner and M. Heimann (eds.). Old-Growth Forests: Function, Fate and Value. Springer, Berlin Heidelberg, Germany.

Sills, E.O., S.S. Atmadja, C. de Sassi, A.E. Duchelle, D.L. Kweka, I.A.P. Resosudarmo and W.D. Sunderlin. 2014. REDD+ on the ground: a case book of subnational initiatives across the globe. Center for International Forestry Research, Bogor, Indonesia.

Silva, L.C.R., M. Anand and M.D. Leithead. 2010. Recent widespread tree growth decline despite increasing atmospheric CO_2. PLoS ONE 5: e11543.

Smith, P., H. Haberl, A. Popp, K.H. Erb, C. Lauk, R. Harper, F.N. Tubiello, A. de Siqueira Pinto, M. Jafari, S. Sohi, O. Masera, H. Böttcher, G. Berndes, M. Bustamente, H. Ahammad, H. Clark, H. Dong, E.A. Elsiddig, C. Mbow, N.H. Ravindranath, C.W. Rice, C. Robledo Abad, A. Romanovskaya, F. Sperling, M. Herrero, J.I. House and S. Rose. 2013. How much land-based greenhouse gas mitigation can be achieved without compromising food security and environmental goals? Glob. Change Biol. 19: 2285–2302.

Smith, P., M. Bustamante, H. Ahammad, H. Clark, H. Dong, E.A. Elsiddig, H. Haberl, R. Harper, J. House, M. Jafari, O. Masera, C. Mbow, N.H. Ravindranath, C.W. Rice, C. Robledo Abad, A. Romanovskaya, F. Sperling and F.N. Tubiello. 2014. Agriculture, Forestry and

Other Land Use (AFOLU). pp. 811–922. *In*: O. Edenhofer, R. Pichs-Madruga, Y. Sokona, E. Farahani, S. Kadner, K. Seyboth, A. Adler, I. Baum, S. Brunner, P. Eickemeier, B. Kriemann, J. Savolainen, S. Schlömer, C. von Stechow, T. Zwickel and J.C. Minx (eds.). Climate Change 2014: Mitigation of Climate Change. Contribution of Working Group III to the Fifth Assessment Report of the Intergovernmental Panel on Climate Change. Cambridge University Press, Cambridge, United Kingdom.

Soil Survey Staff. 1999. Soil Taxonomy: A Basic System of Soil Classification for making and Interpreting Soil Surveys (2nd Ed.). Agriculture Handbook 436. United States Department of Agriculture, Natural Resources Conservation Service, Washington D.C., USA.

Soja, A.J., N.M. Tchebakova, N.H.F. French, M.D. Flannigan, H.H. Shugart, B.J. Stocks, A.I. Sukhinin, E.I. Parfenova, F.S. Chapin III and P.W. Jr. Stackhouse. 2007. Climate-induced boreal forest change: predictions versus current observations. Global Planet. Change 56: 274–296.

Startsev, N.A., V.J. Lieffers and D.H. McNabb. 2007. Effects of feathermoss removal, thinning and fertilization on lodgepole pine growth, soil microclimate and stand nitrogen dynamics. Forest Ecol. Manag. 240: 79–86.

Stephens, B.B., K.R. Gurney, P.P. Tans, C. Sweeney, W. Peters, L. Bruhwiler, P. Ciais, M. Ramonet, P. Bousquet, T. Nakazawa, S. Aoki, T. Machida, G. Inoue, N. Vinnichenko, J. Lloyd, A. Jordan, M. Heimann, O. Shibistova, R.L. Langenfelds, L.P. Steele, R.J. Francey and A.S. Denning. 2007. Weak northern and strong tropical land carbon uptake from vertical profiles of atmospheric CO_2. Science 316: 1732–1735.

Stephenson, N.L., A.J. Das, R. Condit, S.E. Russo, P.J. Baker, N.G. Beckman, D.A. Coomes, E.R. Lines, W.K. Morris, N. Rüger, E. Álvarez, C. Blundo, S. Bunyavejchewin, G. Chuyong, S.J. Davies, Á. Duque, C.N. Ewango, O. Flores, J.F. Franklin, H.R. Grau, Z. Hao, M.E. Harmon, S.P. Hubbell, D. Kenfack, Y. Lin, J.R. Makana, A. Malizia, L.R. Malizia, R.J. Pabst, N. Pongpattananurak, S.H. Su, I.F. Sun, S. Tan, D. Thomas, P.J. van Mantgem, X. Wang, S.K. Wiser and M.A. Zavala. 2014. Rate of tree carbon accumulation increases continuously with tree size. Nature 507: 90–93.

Stocks, B.J. 2004. Forest fires in the boreal zone: climate change and carbon implications. International Forest Fire News 31: 122–131.

Stocks, B.J., B.S. Lee and D.L. Martell. 1996. Some potential carbon budget implications of fire management in the boreal forest. pp. 89–96. *In*: M.J. Apps and D.T. Price (eds.). Forest Ecosystems, Forest Management and the Global Carbon Cycle. Springer-Verlag, Berlin, Germany.

Strassburg, B.B.N., A. Kelly, A. Balmford, R.G. Davies, H.K. Gibbs, A. Lovett, L. Miles, C.D.L. Orme, J. Price, R.K. Turner and A.S.L. Rodriguez. 2010. Global congruence of carbon storage and biodiversity in terrestrial ecosystems. Conserv. Lett. 3: 98–105.

[TFF] Tropical Forest Foundation. 2009. Tropical Forest Foundation Standard for Reduced Impact Logging. Tropical Forest Foundation, Alexandria, Virginia, USA.

Thomas, R.Q., C.D. Canham, K.C. Weathers and C.L. Goodale. 2010. Increased tree carbon storage in response to nitrogen deposition in the US. Nat. Geosci. 3: 13–17.

Thompson, I.D., J. Ferreira, T. Gardner, M. Guariguata, L.P. Koh, K. Okabe, Y. Pan, C.B. Schmitt, J. Tylianakis, J. Barlow, V. Kapos, W.A. Kurz, J.A. Parrotta, M.D. Spalding and N. van Vliet. 2012. Forest biodiversity, carbon and other ecosystem services: relationships and impacts of deforestation and forest degradation. pp. 21–50. *In*: J.A. Parrotta, C. Wildburger and S. Mansourian (eds.). Understanding Relationships between Biodiversity, Carbon, Forests and People: The Key to Achieving REDD+ Objectives. International Union of Forest Research Organizations, Vienna, Austria.

Trumbore, S. 2006. Carbon respired by terrestrial ecosystems—recent progress and challenges. Glob. Change Biol. 12: 141–153.

Turetsky, M.R., M.C. Mack, T. Hollingsworth and J.W. Harden. 2010. Patterns in moss productivity, decomposition, and succession: implications for the resilience of Alaskan ecosystems. Can. J. For. Res. 40: 1237–1264.

Tyrrell, M.L., J. Ross and M. Kelty. 2012. Carbon dynamics in the temperate forest. pp. 77–107. *In*: M.S. Ashton, M.L. Tyrrell, D. Spalding and B. Gentry (eds.). Managing Forest Carbon in a Changing Climate. Springer, Netherlands.

[UNFCCC] United Nations Framework Convention on Climate Change. 2002. Report of the Conference of the Parties on its seventh session, held at Marrakesh from 29 October to 10 November 2001. United Nations Framework Convention on Climate Change. Bonn, Germany. Available at: http://unfccc.int/bodies/body/6383/php/view/reports.php

[UNFCCC] United Nations Framework Convention on Climate Change. 2012. Conference of the Parties serving as the meeting of the Parties to the Kyoto Protocol - Eighth session - Doha, 26 November to 7 December 2012. Report of the Ad Hoc Working Group on Further Commitments for Annex I Parties under the Kyoto Protocol. United Nations Framework Convention on Climate Change. Bonn, Germany. Available at: http://unfccc.int/bodies/body/6383/php/view/reports.php

[UN] United Nations. 2014. New York Declaration on Forests. United Nations, New York, USA.

Van der Werf, G.R., D.C. Morton, R.S. DeFries, J.G.J. Olivier, P.S. Kasibhatla, R.B. Jackson, G.J. Collatz and J.T. Randerson. 2009. CO_2 emissions from forest loss. Nat. Geosci. 2: 737–738.

van Mantgem, P.J., N.L. Stephenson, J.C. Byrne, L.D. Daniels, J.F. Franklin, P.Z. Fulé, M.E. Harmon, A.J. Larson, J.M. Smith, A.H. Taylor and T.T. Veblen. 2009. Widespread increase of tree mortality rates in the western United States. Science 323: 521–524.

Vitousek, P.M. and R.W. Howarth. 1991. Nitrogen limitation on land and in the sea: how can it occur? Biogeochemistry 13: 87–115.

Watson, R.T., I.R. Noble, B. Bolin, N.H. Ravindranath, D.J. Verardo and D.J. Dokken. 2000. Land Use, Land-Use Change and Forestry—Summary for Policymakers. Intergovernmental Panel for Climate Change, Geneva, Switzerland.

White, M.A., P.E. Thornton, S.W. Running and R.R. Nemani. 2000. Parameterization and sensitivity analysis of the BIOME-BGC terrestrial ecosystem model: net primary production controls. Earth Interactions 4: 1–85.

Wirth, C., C.I. Czimczik and E.D. Schulze. 2002. Beyond annual budgets: carbon flux at different temporal scales in fire-prone Siberian Scots pine forests. Tellus 54B: 611–630.

Wollenberg, E., M.L. Tapio-Biström and M. Grieg-Gran. 2012. Climate change mitigation and agriculture: designing projects and policies for smallholder farmers. pp. 3–27. *In*: E. Wollenberg, A. Nihart, M.L. Tapio-Biström and M. Grieg-Gran (eds.). Climate Change Mitigation and Agriculture. Routledge, New York, USA.

Wood, T.E., M.A. Cavaleri and S.C. Reed. 2012. Tropical forest carbon balance in a warmer world: a critical review spanning microbial- to ecosystem-scale processes. Biol. Rev. 87: 912–927.

[WRI] World Resources Institute. 2012. World Resources Institute, Climate Analysis Indicators Tool (WRI, CAIT). 2012. CAIT version 9.0. Washington, DC: World Resources Institute. http://cait.wri.org

Xia, J. and S. Wan. 2008. Global response patterns of terrestrial plant species to nitrogen addition. New Phytol. 179: 428–439.

Zhang, Y., M. Xu, H. Chen and J. Adams. 2009. Global patterns of NPP to GPP ratio derived from MODIS data: effects of ecosystem type, geographical location and climate. Global Ecol. Biogeogr. 18: 280–290.

Zimmermann, M., M.I. Bird, C. Wurster, G. Saiz, I. Goodrick, J. Barta, P. Capek, H. Santruckova and R. Smernik. 2012. Rapid degradation of pyrogenic carbon. Glob. Change Biol. 18: 3306–3316.

CHAPTER 4

Forest Ecosystems and Civilization

An Overview of the Footprint of Modernity in the Exploitation-Conservation Relationship

Anthony Goebel McDermott

ABSTRACT

The main objective of this chapter is to provide an overall assessment of the imprint facilitated by the modern view and some of its concrete expressions (i.e., economic systems) in the profound environmental changes that took place in the forest ecosystems around the globe in the centuries that crowned the expansion of such approach (especially from 16th to 20th). The assessment is made through a contextualized analysis of some of its main indicators and constraints in the overall process of integration of the forest cover into the world system. Regarding indicators, this work seeks to approach the general logic of insertion, both direct and indirect of forest in the world economy first and in the world market later, as well as changes in consumption patterns of forest products, its basic types and their changes over time. With regard to the conditions, special attention is given to what we have called the "utilitarian conservationism". The latter understood as the strategy that

Centro de Investigaciones Históricas de América Central, Escuela de Historia, Universidad de Costa Rica.
E-mail: w.goebel@ucr.ac.cr

sits halfway between the supply and demand for forest products and has promoted or limited the inclusion of forests in the global market in different spatiotemporal contexts.

Forests, Environmental History and Modernity

There is no doubt that forests have been fundamental in the history of humanity. Since the beginning of human society, forests and their products (especially wood and firewood) provided energy, shelter and raw materials for the construction of tools, weapons and utensils which have helped *Homo sapiens* evolve to the top of the hierarchy of living beings. Eventually, humans became the largest transformer of all ecosystems. It is also clear that the pace and intensity in which forests have been exploited throughout history were remarkably dissimilar in different parts of the world. Exploitation has depended on objective factors (the material needs, capital accumulation, the market and the amount of population, among others), attitudes, values and the representations that diverse human groups in different contexts have given to different spatiotemporal forest ecosystems.

Based on the acknowledgment of the inherent complexity of the whole historical process, an analysis of the historical evolution of the multiple dimensions and spaces of interaction between forests and humans throughout the history of mankind and global scale societies, would exceed the generous space provided by the editors of this book. Also, from our perspective, such attempt moves beyond the always problematic balance between the general synthesis and the consequent increase in the scale and the minimum desirable remarks of analytical depth around historiography. This, combined with the inherent complexity of environmental history as a historiographical field of study or discipline, reveals the mutual determination between social and natural systems.

From these limits, the objectives for the development of this chapter are notoriously more modest, however, no less challenging. Therefore, this chapter will focus on modernity as a specific social and historical formation, although modern society has not been globally uniform in terms of values, assumptions, dissemination and scope in the various regions of the planet. It can be considered as a milestone in the strengthening of a dominant vision of the natural world. The latter has been pointed out by Donald Worster who, besides being an undisputed pioneer of environmental history, is also one of the most assiduous defenders of the thesis known as "materialistic culture" which is grounded in a rational, secular, progressive and scientific worldview, characteristic of the modern project. He was based on the "idea that the greatest good in life is to improve the physical condition itself; that is, to achieve greater comfort, greater bodily pleasure, and above all, a higher level of opulence, and that this is a more important goal than ensuring the salvation of the soul, and more important than learning to revere nature or God" (Worster 2006). For Worster, it is this ontological and epistemological change that separates modernity from other previous forms of relationships between humans and the rest of nature. Even though a good number of other forms still survive marginally in our day, the greatest relationship responsible for the global nature of contemporary environmental problems is the former.

Even though there are multiple dimensions that can be analyzed from the imprint of modernity in this area, even in a relatively specific field of environmental history (as it is the space of interaction between human societies and forest ecosystems), we will focus our attention on the limits listed above in three specific processes that happen to be inevitably intertwined.

The first one, which will be discussed in the following section, is the analysis of the general logic insertion, both direct and indirect, of forests in the world economy and on the articulated world market. This will be done with special emphasis on the marketing of forest products and the expansion of modern agriculture, analyzing some of the ecological consequences of both routes of insertion of forest ecosystems in the world system. At this point, we refer to the supply.

A second area to be analyzed is clearly linked to the previous one, but not without its own autonomy. We talk about the consumption of forest products, its basic types and their changes over time. Since these are related, no doubt, with deep cultural, political and socioeconomic changes, their contribution to the pace and intensity of logging is important. Now we are talking about demand.

The last section will focus on an area, that from our perspective is in the middle of the supply and demand for forest products, which is the transverse processes of marketing and consumption of these. We refer to the latter as "utilitarian conservationism". Its role in the inclusion of forests and their products in the global market is clearly dual. On one hand, it limited yet framed the "free" game of supply and demand, constituting a first order conditioning for the commercialization of nature. On the other hand, the mercantilist bias, derived from a reductionist and utilitarian vision centered on the resources rather than the ecosystems, helping ensure that the dominant economic system—a clearly unsustainable one- continued to reproduce, however, it did so in a restricted way to avoid its own excess. This is something that sounds familiar today when many defenders of what we consider impossible, "green capitalism" emerge. In other words, this utilitarian conservationism rather than opposing the modern project is an integral part of it. Hence, its inability to generate the necessary systemic change aimed to prioritize the sustainability of life above all other considerations.

In short, this chapter seeks to bring the reader an overview of the modern understanding of its imprint and some of its concrete expressions (i.e., economic systems) on the deep ecological and environmental transformations that, in the centuries that crowned the expansion of such view (especially from 16th to 20th), took place in forest ecosystems around the globe. Considering our own research experience, the assessment is made through a contextualized analysis of some indicators and constraints, key in the overall process of integration of forest cover of the most diverse world regions in the world system.

Forestry Exploitation and Productive Reorganization: Types and Logics of Forest Ecosystem Involvement in the World Market

Wood has been indispensable throughout most of human history as raw material and energy source for the production of various goods (Williams 2007). The elimination of the forest has brought a drastic change in vegetation cover around the globe as a means to rearrange the biophysical environment for productive purposes. This was considered, in the various human societies that developed after the emergence of agriculture, as something indispensable, either for their material supply, the expansion of the stately rents, or to meet market demands.

Thus, the early insertion of forests in the world economy and then within the articulated world market did not escape this double logic: on one side the *direct insertion* of forest products, especially wood, in the manufacturing of diverse goods, on the other side, the *indirect insertion* of forest ecosystems by removing the forest as a prerequisite for the reorganization of productive purposes. This was particularly oriented to the development of commercial crops in high demand mainly in the advanced capitalist countries (Maddison 1991).

We will try to give the reader an overview of the logic, dynamic, and environmental consequences of both forms of insertion of forest ecosystems in the world system with particular emphasis in the context of consolidation of the global market.

Direct Integration: The Global Trade in Forest Products and Economic Pressure on Forests

In the general European context, the consolidation of a global market was one of the pillars of a broader transition of societies, economies and relationships with the natural world. That transition originated what we could call the Old Regime, a mercenary system in terms of economic aspects and essentially extractive in terms of ecology, that has capitalized modern societies guided by the circulation of capital, labor and resources in a global articulated market place. Under the guidance of this process, which did spread over European colonies and areas of influence, the pressure on forest ecosystems was globally intensified. It was due to the increasing demand for goods directly extracted from forests, particularly wood, that many applications turned into strategic tools that until recently enjoyed the prominent place that today occupies oil (Williams 2007) as energy pillar of the capitalist system (Alvater 2005). At the same time, and in the opposite direction, as clearly indicated by Michael Williams, using the conceptualization of Wallerstein, this process facilitated the "final integration" of nature in the market, as the boundaries of the property and environmental transformations changed. This happened in a similar manner with zonation. In the process, the outer areas of a country eventually became the peripheries of another and even some so called centers became semi-peripheries of these peripheral areas, visibly a dynamic and changing process (Williams 2007).

This integration of forests in the global market came from a conception of nature that favored the exchange value over use value, and where forms of active (central) or passive (marginal) integration took place in the market. Whether domestic or international, those forms of integration shaped to a large extent, the dominant seat that different economies and societies occupied within higher, national or global scale relations between human societies and the natural world. The latter has taken place either as providers or consumers of natural capital itself, or as ecological creditors or debtors respectively (Martinez Alier 1998, 2004).

Wood, a forest product *par excellence,* represented the raw material of many economic activities such as shipbuilding, handicraft making and the manufacture of all kinds of goods. In particular, it served as an energy source in the industrial development of the capitalist centers. Its supply, however, showed a remarkable paradox based on the fact that wood is low cost with extensive supply. In this way, it could not be inserted into long distance marketing networks due to its low profitability. However, its highly strategic nature because of its many applications, in the same reproduction of the economic system, mercenary first and then capitalist, caused the development of wood businesses to transgress the rules of the market. That came along with the idea of consolidating a number of wood marketing networks over long distance even with low profitability. Such characteristics of wood, essential but inexpensive, also prompted dramatic changes in vegetation cover in vast areas and expanded into new business frontiers in remote regions (Williams 2007) as observed in Fig. 1.

Fig. 1. African mahogany logs in West Africa in 1920. Obtained from The Encyclopedia Americana, v. 27, 1920. Public domain.

Note that this enormous pressure from the world market in construction being exerted on forest ecosystems from the marketing of forest products did not take place without positive and negative consequences. The environmental transformations generated by various human societies before European expansion, and especially those related to forests, cannot be avoided either.

The Latin American case is particularly revealing in the newly exposed sense. For authors such as Shawn William Miller, for example, unprocessed nature is not capable of maintaining large human contingents by itself. In that way, Native American companies profoundly changed the ecosystems in which they moved (Miller 2007). In fact, Miller asserts that at the time of the arrival of Europeans there was no ecosystem in the Americas without cultural traces. In this way, the great rainforest imagined especially by Europeans in the 18th and 19th centuries was largely a human artifact

(Miller 2007). The "New World" found by Europeans when they arrived to the Americas was in fact according to him an old world since it was profoundly transformed by the large contingent of humans who had already lived there (Miller 2007). From this perspective, the demographic collapse brought a recovery of native ecosystems, which were in a fragile ecological balance after the profound changes made by indigenous societies. As a result, soils, forests, water sources and wildlife were regenerated to the point that the forest cover in the Americas was higher in the 1800s than in the 1500s (Miller 2007).

Therefore, forests, massively regenerated after the dramatic loss of population marked by the arrival of the Europeans, were again systematically plundered under local population growth and more environmentally predatory activities covered in the belief of abundance and inexhaustibility of resources built in the 18th and 19th centuries, and not since the conquest.

Moreover, there is little doubt that the inclusion of forests in the global market was far more limited compared to the free play of supply and demand for forest products. In this sense, one cannot ignore the contextual specificities of various factors in each country and region that limited or promoted, in different historical moments, the inclusion of commercial forests under their rule. Thus, we cannot overlook elements of undeniable impact on the definition of the changing characteristics and pace of economic exploitation of natural resources. The latter involves the forests themselves, but includes patterns of internal and external consumption, economic pressure on resources, a new business dynamic generated around the marketing of natural goods, the creation of institutions that restrict or promote the transformation of forest cover according to the most varied criteria, and the individual or collective action to defend and build a series of relationships between society and nature, not necessarily based on chrematistic values (Martinez Alier 2004). They focus on the exchange value over use value of the natural world, advocated by the dominant culture of nature where it was worthless before its commercial insertion.

To illustrate this point, we have cases like those of British India in the 19th century epilogue (Fig. 2), where the exploitation and marketing of timber and other forest products, were stimulated if they could represent a considerable income for the weakened state economies barely diversified and committed to primary production in a crisis of its engine product (Sivramkrishna 2009). The other cases, that leave no doubt about the institutional weight factor of the state in the pace and intensity of forest, were the ones of those countries where public policies of dual character were implemented. In these instances, the process took place by combining the stimulus to massive logging in certain regions considered for forestry

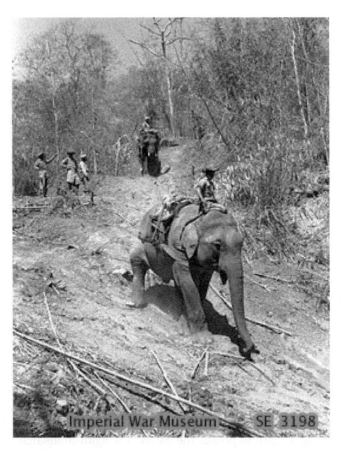

Fig. 2. Men of an Indian forestry company used elephants to transport logs from the timber grounds to the wood yard where they were used to build boats. Obtained from the Imperial War Museums. Public domain.

with selective protection of riparian forests close to rivers and water sources. The United States and Costa Rica are, among others, representatives of such policies (Hays 1999; Goebel Mc Dermott 2013) whose origin and conceptual base is discussed below.

Other countries promoted the massive inclusion of forests in the market taking advantage of specific situations such as the case of Mexico. In the context of World War II, the conflict exerted enormous pressure on the forest cover with the consent of the various governments in the conflagration that saw a tremendous opportunity to insert massive forests into the market. This promoted the indiscriminate cutting of trees (following the abolition of conservation measures established by the state after the end of the revolution in 1917), facilitated the development of commercial forestry, and remained the guide to forest policy even long after the conflict ended (Boyer 2013).

With this, we wanted to show that, while business in timber and other forest products decisively influenced the consolidation of the contemporary world market, the economic exploitation was subject to multiple conditions. Deforestation in short, was part of a wider problem externally related to the expansion of the world market, but also essentially, an internal problem related to the ways in which actors from different societies historically built relationships with their natural surroundings (Wakild 2011a,b; Boyer 2013).

Indirect Insertion: Reorganization of the Biophysical Environment and Agricultural Systems

In the previous section, we tried to size how the marketing of forests and especially of wood products, was established as an activity with its own dynamics. That was focused on the multiple and continuously growing uses that this good acquired over time and on the strategic nature of the use of wood in the reproduction of the economic system itself. In turn, however, wood was clearly a byproduct of the production reorganization of the biophysical environment in regions of the planet that saw commercial agriculture becoming a driver of the international market as well.

Productive transformation of woodlands as well as other processes that affect ecology around the world, have been driven by consumption of European growth markets. This situation has led to an international mobility from the mid 19th century where factors of production were economically more efficient (Foreman-Peck 1995; Fieldhouse 1990). This productive reorganization of the "new economy" was based on trade theory that emphasized the comparative advantage. In this way the income increased with the opening of economies to trade (Foreman-Peck 1995), and some of these economies expanded faster than Europe. Since the latter effect was stimulated by the reduction in freight transport over long distances, such event led to an expansion of the arable land and an increase in the standard of living of the farming population worldwide (Foreman-Peck 1995). Poorly diversified tropical countries experienced monoculture due to the inherent fragility and low productivity of tropical agriculture, which limited the emergence of a modern manufacturing sector (Foreman-Peck 1995) as the essential basis of modern economic development.

Thus, the productive reorganization of the biophysical environment in areas of recent colonization generated both winners and losers in economic terms and was influenced by both endogenous and exogenous factors (Barioch 1981). In its ecological dimension, these attempts to incorporate nature in the market led to an extreme simplification of ecosystems with the introduction and spread of commercial agriculture. This was mediated by the systematic expansion of the agricultural frontier in countries and regions endowed with abundant resources underutilized according to the logic of economic liberalism, although they were distributed unevenly.

As already mentioned, wood was the energy base of industrial growth from its beginnings to the arrival of oil (Fig. 3). Although hydrocarbons freed the economy from its dependence on land as a factor of production, it eliminated the need to conserve part of the forest cover for fuel due to the slow replacement of the wood resource. This situation, encouraged the expansion of modern agriculture and the resulting increase in pace and intensity of deforestation (Wrigley 1993). In other words, when analyzed and interpreted from changes in energy systems, one can say that economic sustainability was constructed from the environmental unsustainability.

Fig. 3. Drawing of peeling hemlock bark for the tannery in Prattsville, New York during the 1840s. At the time, it was the largest tannery in the world. From RH Pease, Wilson Del. Public domain.

This modern agriculture, generated from intense deforestation, brought profound ecological changes since it was more intensive in energy and materials and depended upon more external energy inputs than agroecosystems. As such, it resulted environmentally inefficient despite its overwhelming increase in economic production. This situation can be contrasted with traditional organic base farming, which happened to be relatively closed, less intensive in energy and materials, equipped with high rates of reuse and conditioned by agroclimatic features such as rainfall in the areas where it was carried out. This of course, makes the latter less productive in economic terms, at least in the short term, and more sustainable in environmental terms. In this regard, authors have developed historical environmental analysis from the perspective of social metabolism (González de Molina 2001; Naredo 2000; Guzman et al. 2007; Cussó et al. 2006; Infante Amate 2000, 2012, 2013) who have sized the way the agroclimatic features, profoundly altered by deforestation, and its effects, have become environmental constraints of the structure and distribution of crops in traditional agriculture organic base (González de Molina 2001). To sum up, the indirect inclusion of forests in the global market took place from two combined views of agricultural systems as primary appropriation of nature forms and simultaneous processes. On one hand, the expansion of modern agriculture and on the other hand, the squandering of energy substrate of traditional agriculture.

Thus, we believe that given the agrarian capitalism as a way of insertion in the subsidiaries of the capitalistic world market, there was a transformation process of the relationship between society and nature that broke up in various regions of the planet in relation with traditional forms of nature exploitation-conservation. This process was clearly differentiated in pace, intensity and outcome. In social and environmental terms, it overthrew the relationship of historical and social subjects within their biophysical environment, and especially their communal and collective forms of customary access to natural resources which was necessary for the survival of various social groups and communities. These were instead replaced by an inflexible system of private ownership, with a subsequent prosecution of pre-modern forms of appropriation and use of nature. Such changes are part of the construction of a second nature, specifically a capitalist one, borrowing from the conceptualization provided by O'Connor (2001).

In short, either by the growing demand for raw materials and products derived from the reorganization of the biophysical environment or by the essential character of wood in the capitalist centers, the world market demanded the growing deforestation of areas from increasingly distant regions of the world, exerting an overall pressure on natural resources. Observed from the perspective of ecological economics, the above implied that the capitalist centers expropriated more environment from the peripheral regions and incessantly expanded their footprint on the planet (Martinez Alier 2004). Deforestation, in this context, was a corner stone to commodify the global biophysical environment in spite of being an activity that accompanied the human species from its own origins (Williams 2007). From the second half of the nineteenth century, deforestation had a remarkable acceleration driven by the transit of expensive and exotic tropical agricultural products whose consumption was limited to the affluent sectors of the capitalist centers. Low cost products and mass consumption (Williams 2007), for strategic applications such as construction and energy, transcended all economic considerations and were ensured at all costs (Williams 2007). Therefore, deforestation could be considered as one of the pillars in which the consolidation of the contemporary world market settled. Furthermore, the expansion of the agricultural and commercial frontier driven by changes in consumption patterns in the capitalist centers, led to the most dramatic impacts on the global biota, replacing tropical vegetation with crops now characterized by high demand and irreversibly altering the vegetation cover around the planet (Williams 2007). Thus, although the impact of the processes described acquired a global outlook, there is no doubt that the impact of these changes was particularly dramatic in tropical countries. It was in these areas where these processes consolidated the marginal position of the region in the world market, forever transforming both their societies and their natural environment. Perhaps the analytical nuance that might introduce this conceptualization of Williams, from various studies that have analyzed in depth the different and varied specific logic of inclusion of forests in the commercial dynamic (Tucker 2000; Boyer 2013; Sivramkrishna 2009), is the fact that these inexpensive and mass consumption products, which were brought to market after deforestation of biodiverse forest, were not always agroindustrial products, such as coffee, bananas, tobacco, cocoa or others. These goods were sometimes new forest products that were cultivated or exploited, whose market demand had also dramatically increased.

Regarding the relationship between joint world market and massive destruction of forest cover on a global level, we consider relevant to introduce some interpretive nuances. Indeed, we must clarify that forest destruction, as a prerequisite for productive rearrangement of the natural biophysical environment, and whose main justification was based on the pursuit of material progress, was not exclusive of capitalist dynamics and ethos in which the latter is based. Some valuable examples and existing lax forest policies in socialist countries like China showed that this feature is inherent in the logic of modern projects, regardless of the ideological sign that they take, even though the political system would interfere in the features, pace and intensity of deforesting processes (Jinlong 2009).

In fact, some years ago, Donald Worster reminded us that it was the imprint of modern design and the world view associated with it (materialistic, secular, progressive and rational) responsible for the major changes in the global environment (Worster 2006). Noting also that "Marx and the Marxists were radical in their quest for social justice, but at the same time they firmly believed in the target of material abundance, and were devotees of the modern world view (own translation)".

The fact is that the particular characteristics of the world demand for forest products—as a goal in itself—and the widespread consumption of the "fruits" of deforestation, especially from tropical countries (Williams 2007), were established at a time when indisputable causes and consequences of massive deforestation intensified settling the capitalist system around the globe.

Forest Consumption: A Globally Differentiated Dynamic

As already mentioned, when observed from a global perspective, there is no doubt that the inclusion of commercial forests caused a growing process of massification or "commoditization" of the resource. However, what were some of the major luxury and mass uses of timber taken from or planted in the most distant regions of the world? How were they changed? To provide some answers to these questions, we will briefly describe some of the main patterns of wood consumption, the logic of its application, and the relationship of this resource to the changes in the forest cover in different regions of the planet.

We should note, first, that the uses of wood, as already pointed out, were diversifying and growing over time. There is no doubt that the construction of ships became one of the major uses of the resource in the process of consolidation of the European world economy in the 16th century and even in the articulation of the world market by the end of the 19th century. Its importance was such that it contributed to the creation of networks of long distance trade since building wooden ships became one of the vital factors for expanding trade throughout the globe. In fact, between the late 19th century and the early 20th century, large parts of the structure of the boat such as the keel, stem and sternpost were usually made of oak, teak or mahogany while the smaller pieces were made of Spanish cedar, cottonwood and other light wood (Roldan 1864).

Other traditional uses of wood were related to the war effort. Authors such as J.R. McNeill have explored global environmental transformations that have historically generated wars including military uses of wood, many of them introduced after the so called military revolution between the years 1450 and 1700. At that point, fortifications in various parts of the world, mainly built by the imperial powers both in the metropolis and its colonies, began to require increasingly greater amounts of a steady supply of wood. Such material was needed for the production of artillery, ammunition boxes, barrels of gunpowder, sheds, palisades and boats, among many other uses (McNeill 2004). Those activities significantly increased the pressure on forest ecosystems, sometimes close to the area where wood was needed and other times notoriously distant from these locations.

Wood, however, was also the subject of a clearly conspicuous consumption, carried out by the more affluent sectors of society. Such is the case of timber typically obtained from the extractive exploitation of wood from tropical forests such as Spanish cedar, mahogany and cocobolo (Fig. 4). These fine woods were used in the manufacture of furniture, crafts, game objects such as chess pieces, tool handles, musical and scientific instruments, piano boxes, veneers and plywood among others (Jiménez Madrigal 1998). These manufactured goods also give us a clear idea of the industrial and manufacturing activities in the centers of world capitalism that generated extensive destruction of Neotropical forests at the time.

Fig. 4. Spanish cedar tree (*Cedrela* sp.) in the forests of Southern Brazil in 1920. Obtained from The Encyclopedia Americana, v. 27, 1920. Public domain.

Because of its versatility and low cost, balsa stood out in the early incursion into the international market. Balsa was grown in tropical America not in Europe and the United States. It was widely desired because of its versatility and the possibilities for the construction of military supplies as well as making toys, models and many other applications in which its intrinsic features, such as its low weight, resulted in an undeniable advantage over other materials.

The consumption of wood was not only differentiated from its uses but also by the level of consumption and the consequent expansion of the forest footprint by different countries and regions. Perhaps the most remarkable aspect is how historically the United States gradually became a growing importer of forest products, displacing Europe and other regions of the planet by the second half of the 20th century. Thus, the United States was consolidated at the forefront of both precious wood imports, intended for conspicuous consumption of the dominant sectors, as well as "cheap" consumer wood. This country also promoted the development of plantations of fast growing trees such as balsa, especially for the manufacture of military equipment in a context of war economy (Tucker 2000; Boyer 2013). The latter caused the usual ecological and social consequences resulting from the substitution of the biodiverse forest for one highly simplified, such as the loss of ecological functions essential for sustaining human life (Martinez Alier 2004). At this point, it is important to highlight that this process of massification of forest products took place in a context in which the United States forest industry had accumulated enough financial and technological capital to risk investing in forestry projects in tropical regions (Tucker 2000) which led undoubtedly a growing process of industrialization and transnationalization of the timber industry.

This leadership was closely linked with the increase and diversification of consumption of forest products in the United States, which had made it a net importer of these products after being a historically net exporter (Williams 2006). This picture was in contrast with the data by the United States in the early 20th century when, according to the meticulous work of Zon and the Research Section of the United States Forest Service in the first calculation of global forest resources, the production-consumption relationship of forest products showed a massive deficit in the developed world. While the production was 845.4 million cubic meters, the consumption equaled 1,312.4 million cubic meters (Williams 2006). This created a deficit of –467.0 million cubic meters of wood, of which an impressive –638.7 (above the overall deficit of the developed world) corresponded to the United States, which as noted, expanded its forest footprint globally. This is also evident by the fact that in this same period the United States concentrated 15% of world imports of timber (Williams 2006).

In a thorough analysis of United States imports in the mid 1950s, Don D. Humphrey highlighted the fact that forest products accounted for 10% of total imports, while the wood gross imports had increased from 4.4% of national production in 1924 to a maximum of 8.7% in 1950. Thus, by 1950 the United States imported more than a trillion dollars in paper and wood as a trend for sustained growth (Williams 2006). In fact, according to this author, the growth rate of the local timber production in the United States, in terms of cubic feet, was approximately equal to the rate of depletion of forest reserves. Yet, the depletion of forests was greater considering the relationship between cutting rates and growth since much of the "leakage" of forest resources was concentrated in lumber, especially conifers, while most of the growth was in low grade trees. Then, the growing inability of local timber supply to meet a growing demand for forest products was the central cause of the trend in US imports, especially in the immediate post World War II period. This undoubtedly created a growing pressure on forest cover in different areas of the planet.

Given a clear tendency of the US market and other wood markets to mass import wood that they could not locally produce, and consequently expand its forest footprint taking over an increasing portion of the forest capital of other world regions, it is worth asking which industry or industries exerted greater pressure on forests globally? Or in other words, is it possible to measure the qualitative and quantitative changes in both supply and demand for forest products? We observed the particularly use of balsawood in the construction of various military equipment and the requirements of the Department of State that stimulated the development of balsa plantations in countries like Ecuador and Costa Rica. However, again, the work of Humphrey (1955) provides important additional evidence about less cyclical changes in consumption patterns of forest products in the US allowing us to, at least partially and provisionally, answer these questions. According to that author, the publishing market, more specifically the one related to the press, was one of the sectors of the economy whose growing demand triggered the aggregate demand for forest products. This industry, which required huge amounts of wood pulp for paper making, had become, in the early 1950s, the sector of the economy that required the importation of forest products, even though it once obtained the pulp from local sources. Thus, while in 1929 the United States had imported a total of 4845 million pounds of newsprint, imports in 1949 amounted to 6634 million pounds (Humphrey 1955). Despite this, the local production of this sector was increasing, so imports of wood pulp grew in the early fifties at a lower rate than other forest products, especially those related to construction. Indeed, from the 1940s, the United States began to import logs and sawn wood at a relatively low cost, which was related, among other factors, with the boom in housing construction and timber requirements for the construction of military equipment, as well as plywood, used for both industrial and domestic purposes. This qualitative diversification and quantitative increase of the demand for wood in the U.S. was related, in its characteristics and in its pace and intensity, with changes in the amount and pace of exports of wood of different countries and regions of the world. Those observations took place even in those forests whose wood had virtually no commercial value before this process of "commoditization" of the resource (Tucker 2000).

In addition, new and growing uses of timber in the main importing countries of forest products, profoundly affected the logic of forestry exploitation in various regions of the planet. Observed from the perspective of the World System from Immanuel Wallerstein and the specific forest dimension of an outdoor area or semi periphery to a classic periphery (Wallerstein 1989), these uses of wood were gradually changing. As mentioned before, wood switched from being used to producing expensive and exotic goods, whose consumption was limited to the affluent capitalist centers, to low cost products and mass consumption (Williams 2007). The latter was a decisive factor in accelerating deforestation processes around the globe with dramatic consequences, particularly in tropical countries.

The United States market, however, was not alone in leading the economic exploitation of certain species. In fact, as we have seen, its forest footprint was especially expanded at the dawn of the XX century. European countries like Germany, France and Britain, among others, had a long tradition of forestry exploitation of precious wood from the tropics, especially for the manufacture of fine furniture, musical instruments and other luxury goods. Although in a context of scarce growing timber this dominated the stage of commercialization of such good. From the late 19th and early 20th century tropical forests were just beginning to be considered a source of timber, regardless of type (Williams 2007). This observation seems to confirm the growing massification process faced by forest products in general, which made an enormous pressure and global nature of the forest cover on the planet and not just on regions with forests where precious woods were extracted.

This pressure was also expressed in the growing import of low value timber such as the cashew tree, known as *espavel* in Spanish, even in countries such as Peru, which imported huge amounts of this and other wood species from Central America (Goebel 2014) that in the late 19th century were just beginning their transition toward capitalistic modernization. Although the picture was not entirely clear in these cases, it is true that some authors have reported the existence of vast wood marketing networks on both sides of the Pacific Ocean. In those, the "cheap" massively exploited Oregon woods, met the needs of building cities and countries that to the late 19th century were not self-sufficient in wood for construction. An example is the case of Victoria, Australia, where redwoods were used in the manufacture of sleepers for the railway, especially after the excesses that followed the gold rush (Williams 1997). Within these Pacific countries which presented deficits in timber "production", Williams highlighted the fact that Ecuador and Peru proved to be lucrative markets for timber companies settled along the coast of the Pacific Ocean. This deficit and sub sequential dependency on foreign timber accelerated deforestation that led to the introduction and spread of various economic activities such as mining, agriculture and livestock export, present in the other countries of the Pacific Rim, with the notable exception of New Zealand, a country where a booming timber industry developed (Williams 1997). Not surprisingly, cheap wood that is both strong and useful, particularly for work in construction, such as cashew tree, were exported to countries like Peru.

The "Utilitarian" Conservationism: A Historic Continuum

Despite that the transformation of nature, and particularly forests, into natural goods of production and consumption had as a key element the discounting of functional ecosystem interactions from the economic calculations (Worster 1996), many critical voices sensed that the unquestionable and nearly dogmatized logic of "progress" had a dark side. In the Latin American Caribbean and other imperial colonies, for example, the first display of monoculture took place in the XVIII century with the development of sugar, cocoa, coffee and tobacco plantations. The radical simplification of the ecosystems inherent to these changes did not only bring along important environmental consequences, but also founded the bases of modern conservationism according to some authors.

Several of these measures were based on the established or perceived relationship between deforestation and the depletion of the river flows and water sources. Hence, according to Richard H. Grove, quoted by Woster (1996), the effect of monoculture over the future profitability of the soil in the French colonies of the Indian Ocean prompted the perception of a direct kind between "the deforestation and the local climate change". It did not take long before the English followed the French example, when naturalists such as Stephen Hales and SoameJenyns "established a causal relationship between the trees and rain, thus identifying the dangers of deforestation and their impact over soil erosion".

The latter lead to the establishment, in 1764, of forest reserves in Tobago identified in maps as "forest reserves for the rains", which comprised according to Grove, nearly 20 percent of the territory. Very distant from that, in the Bourbon Spain, the principle of "mount conservationism" was reflected in the legislation of this imperial power, radiating itself with varied levels of implementation, towards colonies they aimed to control (Morera Jiménez 2006).

In the emergence of the first forest protection measures based on their relation to the water resource, we cannot leave aside the contribution of Humboldtian science. In fact, most of the aggressive forest conservation programs and promotion of forestry that were developed especially during the 1930s and 1940s in different parts of the planet were great contributors to the concepts about the direct relationship between deforestation and climate change developed by the Prussian scientist. The latter took place in areas of the Soviet Union, the United Stated and Nazi Germany and several Latin American countries such as Venezuela and Peru (Cushman 2011). The evidence for the ideas came from the reduction of the water level of the rivers, lakes and springs in what was called the "droughtiness theories", despite the fact that Humboldt himself would later attribute part of these processes to the influx of specific climatological events.

These policies, laws and practices inherent to the early conservationism that aimed to limit the systematic exploitation of nature, represented among other elements, an acknowledgment to the limited nature of resources. Therefore, the term "utilitarian" that we have attributed to such conservation logic is based on the following consideration. What finally prompted the enactment of laws, the establishment of policies and the implementation of measures that would limit the unrestricted exploitation of nature, with special emphasis on the forests as water producer, was certainly a consequence of the fear that capitalism as an economic system would stop reproducing. In other words, in the framework of a dominant vision of nature guided by the "progress" philosophy, the conservationism already established in the epilogue of the 19th century, brought the attention to the dangers of the accelerated pace, the excessive intensity and speed of the transformat of nature.

The simplified vision inherent to the conservation of a limited quantity of strategic resources essential for the sustainable growth of economy and not for the sustainability of life on the planet, does not only have strong environmental consequences. This approach, at the same time has already produced a noticeable invisibility of concepts, theories and forms of knowledge, whose applicability implied questioning and even redefining the premises of the modern ethos in general, and of capitalism in particular, leaving without doubt, the socially constructed character of the scientific knowledge (Ortega Ponce y Arrellano Hernández 2010; Dagnino et al. 1998; López Cerezo 1999). This seems to be evident in the development of actual ecological concepts derived from the systematization in Europe of methods for scientific forest management and from studies about the efficient use of energy and agricultural chemistry (Martínez Alier 2004). For instance, the development of theories such as the greenhouse effect idea by Svante Arrhenius, even though it appeared between the middle and the end of the 19th century, had a slow promotion and popularization out of the academic intellectual community. This idea was also localized, and it did not represent an obstacle for the "wheels of progress" to continue transforming the natural ecosystems around the globe. The latter was especially prevalent in regions where the pristine myth (Denevan 1992) of an abundant and inexhaustible nature came along with the European exploration; a process called Euroimperialism by Mary Louise Pratt (Pratt 1992). This vision, clearly minimizing the natural environment, defines the value of nature on its commercial insertion, including the cost of its appropriation and exploitation. As such, it is substantially low if not non-existent in economic terms (Martínez Alier 2004) particularly in poor countries as we still see it today.

Towards the end of the 19th and beginning of the 20th century, this "utilitarian conservationism", based on a minimizing and mechanistic top-down vision of nature was accepted especially in the advanced capitalistic countries. The model replicated itself, with clearly contextual variants and nuances in the colonies or sphere of influence of Europe and the United States. At the same time, the main concepts of this epistemology of conservation gave rise to two main trends of self-conscious and organized conservationism, which are differentiated and at the same time unfailingly related.

Based on the conceptualization made by Martínez Alier (2004), the first of these trends, the one this author calls "the cult for wildlife", which promoted the defense of pristine nature and the love towards forests and rivers. This approach aimed to keep the remaining nature pristine and out of the market for many different reasons among which we could mention the scientific appraisal, the aesthetic admiration of the natural landscape, and even the expectation of future usages of the resources. In this way, the idea did not attack the economic growth directly. This trend had John Muir as its first representative and the Sierra Club of the United States towards the end of the XIX century (Fig. 5). Despite its variants and the emphasis that have given origin to diverse subdivisions, the idea still subsists today. In this approach, the priority was placed on the preservation of species, mainly the endangered ones, above any other possibility of commercial use of nature. This conservationist trend was based in the diffusion of the so called pristine myth. It started from the premise that society and nature were two types of ontologically different organizations. This approach obviously left aside the "natural" dimension of the human being, which is our condition as a species, as well as the social dimension of nature where

Fig. 5. United States president Theodore Roosevelt (left) and nature preservationist John Muir (right), founder of the Sierra Club, on Glacier Point in Yosemite National Park during 1906. Obtained from the United States Library of Congress. Public Domain.

either at the material or symbolic level, virtually the whole natural world had been appropriated by society (Cronon 1996) as observed in Fig. 6.

That perspective started from the wrong premise that nature existed outside the sphere of human action and that it had to be "protected" with the implication of some sort of "permit" to continue constantly transforming the biophysical environment. In spite of the fact that nature had already been appropriated and transformed, especially in cities, this preservationist trend was promoted around the planet. In fact, a number of national parks themselves, created during the time of promotion of this idea, are social constructions because many of these "pristine" landscapes were created by governments, institutions and private entities moving aside other forms of preexisting settlements (Gregg 2010). These efforts

Fig. 6. A tiger hunt in a tropical forest with the hunting party riding animals. Etching by J. W. Lowry after W. Daniell. Obtained from the Wellcome Library. Copyrighted work available under Creative Commons Attribution only license CC BY 4.0.

were aimed to "create" national parks by planting trees and transplanting biodiversity in order to build an idealized nature without humans but without risks for the visitors who, around the world, accessed even more frequently these spaces, immersed in the nostalgia of finding a "natural" world farther away over time. The national parks and protected areas are social constructions generated starting from a wrong idea of "nature". More than pristine natural spaces, they are and have largely been true conservation infrastructures (Carse 2012; White 1995 and Pritchard 2011). They constitute clear expressions and even clearer indicators of the general unsustainability of the societies which have "created" them. Then, it is not strange that the touristic promotion, especially the North American adventure tourism, found a great deal of their historical roots in this trend of worldwide conservationism and especially in the pristine and exotic character attributed to nature. Adventure tourism in pristine and "rough" landscapes was also a way to reinforce the masculinity of the North American members, who saw in this practice the chance to get close to their primitive roots and scape, at least briefly, from the "feminizing" tendencies of the civilized life (Cronon 1996; Rome 2006). That way, as well stated by the environmental historian William Cronon (1996), this nostalgic search of the lost frontier became, in the American case, in one important vehicle for the expression of antimodernism.

A second stream of conservationism whose influence is still prevalent today, is what Martinez Alier (2004) has called the "gospel of eco-efficiency", a stream that has focused on the effects of economic growth, not only on pristine nature but on industrialization, agriculture and urban planning. Rather than finding a solution and preserving "what is left" from the natural world (so immaculate and untouchable), this stream worries about "the impact on the production of goods and the sustainable management of natural resources, and therefore in the loss of the attractions of nature or intrinsic values" (Martinez Alier 2004). Its key concepts are among others, "sustainable development" and "ecological modernization", while terms such as "natural resources", "natural capital" or "environmental services" have replaced almost entirely the word "nature".

This approach is based on the economic concept of sustainable development, and with the strong, arrogant, and anthropocentric belief that science could determine biological growth rates. In this way, regular crops are expected to comply with these rates and indicate "precisely how many trees they could be used without reducing the forest itself or affect its biological capacity for the long term" (Worster 2006). This conservation stream, conceptualized by Samuel P. Hays (1999) more than fifty years ago as the Progressive Conservation Movement, had Bernhard Fernow and Gifford Pinchot among its main representatives. The second being the first director of the United States Forest Service, and the person who imported these concepts from the development of European forestry.

In this simplified and visibly reductionist view of nature, especially of forest ecosystems "nature was a little more than a utility to be managed and harvested for the common good" as stated by Worster (2006):

> *"They had completely made theirs the dominant world view at the time, for which economic progress—the steady increase in long-term production-was the primary goal of social life, adding only that production should be directed by the state and its experts to prevent the destruction of the organic social order".*

The origin and evolution of this "progressive" conservatism in the United States have been analyzed in depth by Hays (1999). To summarize, we can say that he realized, among other aspects, the clashes and disputes between defenders and diverse social, political and economic groups on the different views of ownership, use, and conservation of nature components. After all, these conflicts do not focus on nature itself, finally conceived as a single resource of multiple uses. Also, defenders of this strategy of "conservation", characterized by a strong state presence for the "development", management, and exploitation of natural resources, had the strong opposition of the counterpart exemplified by aforementioned defenders of "cult" wildlife, among which certainly stood the figure of John Muir and the Sierra Club.

The progressive trend has evolved historically and is widely accepted today although it has two main weaknesses. First, the difficulty of establishing what is "proper use of resources" or how to reach, from the perspective of a green economy, a strong sustainability. Second, the recognition of the need to exploit resources not that "irrationally" has become a powerful justification to continue the systematic depletion of natural resources, using the metaphorical cloak of a poorly defined sustainability. In this way, the main weakness is the risk of perpetuating the "modern" and unsustainable development paradigm.

In short, this stream of conservation advocated the need of nature to be at the service of economic progress but in a regulated manner. It needed to have a strong control from the government about certain strategic resources, and avoid the predatory nature of the inherently short-termism trade, exhausting the natural capital especially represented by forests. There were many countries where such control and regulation of forestry exploitation implied that the forests and their productions were simultaneously "protected" from the spontaneous exploitation by communities, peasants, indigenous and other groups increasingly deprived of the ecological distribution. However, these groups exploited the forests outside the market as well as the scientific rationality and other canons of western modernity in relation to nature. This implied a serious waste of resources even when many of these social groups were notoriously sustainable in terms of pollution.

In other words, the utilitarian conservationism was finally driven not only by a strictly economic value of nature. At the same time, given their modernist and mercantilist bias, it was and still is socially exclusive for ignoring any other type of assessment and relationship with the natural world which is different from the economic and scientific rationality in the management of resources for "progress".

Thus, both exploitation and conservation of forests, which could be considered as the "two faces" of the same modern project, generated various ecological conflicts (Martinez Alier 2004; Folchi 2001 and González de Molina 2009) from individual or state appropriation and exploitation of the forests at the expense of other collective, customary, and traditional forms of access to forests and their productions. Just to mention an example, in the Spanish case, the processes of commodification of public forest carried by the Spanish state in the mid 19th century and early 20th generated forms of peasant resistance to the elimination of the traditional free access to resources (Sabio 2002). As clearly stated by Alberto Sabio, "against the productivist orientation and commodification of forest officials, some peasant struggles for keeping their access to natural resources would contribute to reduce environmental degradation by keeping those resources safe from the commodification and from commercial exploitation". This did not mean "that the poorest were ecologically innocent, but that separating natural resources from the market, they were closer to a green economy, some financial considerations and sometimes near the Thompsonian expression of moral economy" (Sabio 2002).

The conflict generated by this clearly asymmetrical appropriation of other resources such as water was more frequent in various contexts in Latin America and elsewhere, between the "progressive technocrats" and peasants, indigenous groups and other individuals or social groups. This was particularly evident after the simultaneous enactment of conservation laws, which were economically "efficient" but socially exclusive and sustainably questionable in terms of pollution. This happened, for example, in Peru, Mexico and Russia (Cushman 2014) just to mention a few cases where conflicts over water resources had a significant social, political and economic impact.

In the case of other resources such as wood, there were frequent disputes in regions that have not yet completed its energy transition and where firewood is still their basic form of energy. The Latin American case is a good representative of such conflicts particularly since the late 19th century (Tortolero Villaseñor 2009; Lobato Martins 2011).

The visible modern and mercantilist bias and the socially exclusionary character of the two streams of utilitarian conservationism outlined should not lead to the mistake of thinking that finally, its application, clearly differentiated in the various regions of the world, had no positive effects on the natural biophysical environment. It is difficult to size its real impact and leads inevitably to pose counterfactual assumptions about how they have expanded the modern project around the globe without obstacles. However, we will take the risk of posing here, as a hypothesis, that the utilitarian conservationism may have earlier impacted the threshold of sustainability of life on the planet. This included biodiversity loss, reduced ecological functions of the forest, some alterations in food webs, and human appropriation of net primary production of biomass, only to name a few biophysical indicators of unsustainability. Even though some insist that this threshold has not yet been trespassed, and even that being so, there is nothing science and technology can do to ensure our way of life and our current consumption levels, the process of change has certainly exerted an accelerated pressure on the limits of resources.

In short, this utilitarian conservationism, which often tends to be mistakenly confused with environmentalism that came on the scene after the second half of the 20th century, markedly anti-systemic and settled on a real ecological paradigm, had positive consequences for the environment. In the case of forests, for example, whole ecosystems were protected when attempting to protect a particular resource from the "excesses" of the market but also, the protection of endangered species and altered food webs, and the limited or prohibited use of riparian forests impacted the pace and intensity of global climate change. These and many other consequences of the global implementation of laws, policies, and conservation strategies developed from the perspective of this type of conservationism, and as it is common in many areas of human activity, a correct result was achieved starting from a wrong premise.

Acknowledgements

This work is not an isolated research project that can be disclaimed from the historiographical project it belongs to. That is why I want to deeply thank the Historical Research Center of Central America (CIHAC) of the University of Costa Rica, and especially to its current director Ph.D. Juan José Marín Hernández, not only for encouraging historical studies of the environment but also for including me in its select team of researchers. Specifically, this chapter received multiple inputs from research projects developed by the writer along with other colleagues, covered in the research program No. 806-A6-911 Regional Comparative History of Costa Rica, Central America and the Caribbean, CIHAC. I also want to thank Ph.D. Ronny Viales Hurtado for his always valuable comments and suggestions on how to approach a synthesis such as this. Finally I want to thank the editors of this book, Ph.D. Sergio A. Molina-Murillo and Ph.D. Carlos Rojas, for taking me into account to participate in this academic adventure, for their suggestions and guidance for the realization of this chapter. Errors of form and content are exclusive responsibility of the writer.

References

Alvater, E. 2005. La ecología de la economía global (I). *In*: Diálogos, La insignia, diciembre de 2005, available at: http://www.lainsignia.org/2005/diciembre/dial_002.htm

Alvater, E. 2004. La ecología de la economía global o el ascenso y ocaso del régimen de energía fósil. pp. 17–52. *In*: J. Ponce (ed.). Globalización: la euforia llegó a su fin. Ediciones Abya-Yala, Quito, Ecuador.

Bairoch, P. 1981. Las grandes tendencias de las disparidades económicas nacionales después de la Revolución Industrial. pp. 196–213. *In*: J. Topolsky, J. Cipolla, C. Bairoch, P. Hobsbawm and E.J. Kindleberger (eds.). Historia económica: nuevos enfoques y nuevos problemas. Crítica, Barcelona, Spain.

Boyer, C. 2013. La Segunda Guerra Mundial y la 'crisis de producción' en los bosques mexicanos. HALAC 2: 7–23.

Carse, A. 2012. Nature as infrastructure: making and managing the Panama Canal watershed. Social Studies of Science 42: 539–563.

Cronon, W. 1996. The trouble with wilderness: or, getting back to the wrong nature. Environ. Hist. 1: 7–28.

Cushman, G.T. 2011. Humboldtian science, creole meteorology, and the discovery of human-caused climate change in South America. Osiris 26: 19–23.

Cushman, G.T. 2014. Guano & the Opening of the Pacific World: A Global Ecological History. Cambridge University Press, New York, USA.

Cussó, X., R. Garrabou and E. Tello. 2006. Social metabolism in an agrarian region of Catalonia (Spain) in 1860–1870: flows, energy balance and land use. Ecol. Econ. 58: 49–65.

Dagnino, R., H. Thomas and E. Gomes. 1998. Elementos para un "estado del arte" de los estudios en Ciencia, Tecnología y Sociedad en América Latina. Redes 11: 231–255.

Denevan, W. 1992. The pristine myth: the landscape of the Americas in 1492. Ann. Assoc. Am. Geogr. 82: 369–385.

Fieldhouse, D. and J.A. Ruiz de Elvira Prieto. 1990. Economía e imperio: la expansión de Europa (1830–1914) (3rd ed.). Siglo XXI, Madrid, Spain.

Folchi, M. 2001. Conflictos de contenido ambiental y ecologismo de los pobres: no siempre pobres, ni siempre ecologistas. Ecología política 22: 79–100.

Foreman-Peck, J. 1995. Historia económica mundial, las relaciones económicas internacionales desde 1850 (2nd ed.). Prentice Hall, Madrid, Spain.

Goebel Mc Dermott, A. 2013. Los bosques del "progreso". Explotación forestal y régimen ambiental en Costa Rica: 1883–1955. Editorial Nuevas Perspectivas, San José, Costa Rica.

Goebel Mc Dermott, A. 2014. Biodiversidad exportada y regiones transformadas: naturaleza, comercio y dinámica regional en Costa Rica (1884–1948). HALAC 2: 339–377.

González de Molina, M. 2001. Condicionamientos ambientales del crecimiento agrario español (Siglos XIX y XX), pp. 43–94. In: J. Pujol, M. González de Molina, L. Fernandez-Prieto, D. Gallego and R. Garrabou (eds.). El pozo de todos los males: Sobre el atraso en la agricultura española contemporánea. Crítica, Barcelona, Spain.

González de Molina, M. 2009. Sociedad, naturaleza, metabolismo social. Sobre el estatus teórico de la historia ambiental. In: R. Loreto López (ed.). Agua, poder urbano y metabolismo social. Benemérita Universidad Autónoma de Puebla, Puebla, México, 245pp.

Gregg, S. 2010. Managing the Mountains: Land Use Planning, the New Deal, and the Creation of a Federal Landscape in Appalachia. Yale University Press, Connecticut, USA.

Guzmán Casado, G.I. and M. González de Molina. 2007. Agricultura tradicional versus agricultura ecológica. El coste territorial de la sustentabilidad. Agroecología 2: 7–19.

Hays, S.P. 1999. Conservation and the Gospel of Efficiency: The Progressive Conservation Movement, 1890–1920. University of Pittsburg Press, Pittsburg, USA.

Humphrey, D. 1955. American Imports. The Twentieth Century Fund, New York, USA.

Infante Amate, J. 2012. Cuántos siglos de aceituna. El carácter de la expansión olivarera en el sur de España (1750–1900). Historia Agraria 58: 39–72.

Infante Amate, J., D. Soto Fernández, A. Cid Escudero, G. Guzmán Casado, P. de Olavide and M. González de Molina. 2013. Nuevas interpretaciones sobre el papel del olivar en la evolución agraria española. La gran transformación del sector (1880–2010). Oral presentation at the XIV Congreso Internacional de Historia Agraria, Badajoz, Spain. November 7–9.

Infante Amate, J. y M. González de Molina. 2000. 'Sustainable de-growth' in agriculture and food: an agro-ecological perspective on Spain's agri-food system (year 2000). J. Clean. Prod. 38: 27–35.

Jiménez Madrigal, Q. 1998. Árboles Maderables en peligro de extinción en Costa Rica, 2. Ed. Instituto Nacional de Biodiversidad de CR, Heredia, Costa Rica.

Jinlong, L. 2009. Reconstructing the History of Forestry in Northwestern China, 1949–1998. Global Environment 3: 190–221.

Lobato Martins, M. 2011. A política florestal, os négocios de lenhae o desmatamento: Minas Gerais, 1890–1950. HALAC 1: 29–54.

López Cerezo, J.A. 1999. Los estudios de ciencia, tecnología y sociedad. Revista Iberoamericana 20: 217–225.

Maddison, A. 1991. Historia del desarrollo capitalista. Sus fuerzas dinámicas: una visión comparada a largo plazo. Editorial Ariel, Barcelona, Spain.

Martínez Alier, J. 1998. Deuda ecológica vs. Deuda externa. Una perspectiva latinoamericana. Parlamento Latinoamericano. *In*: A. Acosta (ed.). Un continente contra la deuda—perspectivas y enfoques para la acción. 383pp.

Martínez Alier, J. 2004. El ecologismo de los pobres. Conflictos ambientales y lenguajes de valoración. Icaria Antrazyt-Flacso, Barcelona, Spain.

McNeill, J.R. 2004. Woods and warfare in world history. Environ. Hist. 9: 388–410.

Miller, S.W. 2007. An Environmental History of Latin America. Cambridge University Press, New York, USA.

Morera Jiménez, M. 2006. Los orígenes del discurso conservacionista en Costa Rica: Estudio de caso Heredia (1821–1840). M.S. Thesis, Universidad de Costa Rica, San José, Costa Rica.

Naredo, J.M. 2000. La modernización de la agricultura Española y sus repercusiones ecológicas. pp. 55–86. *In*: M. González de Molina and J. Martínez Alier (eds.). Naturaleza transformada, estudios de historia ambiental en España. Icaria editorial, Barcelona, Spain.

O'Connor, J. 2001. ¿Qué es la historia ambiental? ¿Para qué historia ambiental? *In*: J. O'Connor (ed.). Causas Naturales: ensayos de marxismo ecológico. Editorial Siglo XXI, México. 403pp.

Ortega Ponce, C. and A. Arellano Hernández. 2010. Relaciones sociales y de genes: el primer vegetal transgénico mexicano. Universidad Autónoma del Estado de México, Mexico.

Pratt, M.L. 1992. Imperial eyes: travel writing and transculturation. Routledge, London, United Kingdom.

Pritchard, S.B. 2011. Confluence: The Nature of Technology and the Remaking of the Rhône. Harvard University Press, Cambridge, Massachusetts.

Roldan, M. 1864. Cartilla de construcción y manejo de los buques para instrucción de los guardias marinas. Imprenta de la Revista Médica, Cádiz, Spain. (Corrected version by Francisco Chacón y Orta. Original from 1831). Available at: https://play.google.com/store/books/details?id=TZ1WAAAAcAAJ&rdid=book-TZ1WAAAAcAAJ&rdot=1.

Rome, A. 2006. Political hermaphrodites: gender and environmental reform in progressive America. Environ. Hist. 11: 440–463.

Sabio, A. 2002. Imágenes del monte público, "patriotismo forestal español" y resistencias campesinas, 1855–1930. Ayer 46: 123–154.

Sivramkrishna, S. 2009. Macroeconomic and environmental history: the impact of currency depreciation on forests in British India, 1873–1893. Global Environment 4: 118–155.

Tortolero Villaseñor, A. 2009. ¿Anarquistas, ambientalistas o revolucionarios? La conflictividad rural en Chalco. San Francisco Acuautla contra Zoquiapa, 1850–1868. Revista de Historia EUNA/EUCR 59-60: 15–34.

Tucker, R.P. 2000. Insatiable Appetite. The United States and the Ecological Degradation of the Tropical World. University of California Press, Berkley, USA.

Wakild, E. 2011. Revolutionary Parks: Conservation, Social Justice, and Mexico's National Parks, 1910–1940. University of Arizona Press, Arizona, USA.

Wallerstein, I. 1989. El moderno sistema mundial I. La agricultura capitalista y los orígenes de la economía-mundo europea en el siglo XVI. Siglo XXI editores, Mexico.

White, R. 1995. The Organic Machine: The Remaking of the Columbia River. Hill and Wang, New York.

Williams, M. 1997. Ecology, imperialism and deforestation. *In*: T. Griffiths and L. Robin (eds.). Ecology and Empire: Environmental History of Settler Societies. Keele University Press, Stafford, United Kingdom.

Williams, M. 2006. Deforesting the Earth: From Prehistory to Global Crisis: An Abridgment. The University of Chicago Press, Chicago, USA.

Williams, M. 2007. The Role of Deforestation in Earth and World-System Integration. *In*: A. Hornborg, J.R. McNeill and J. Martínez-Alier (eds.). Rethinking Environmental History. World-System History and Environmental Change. Altamira Press, Lanham, USA.

Worster, D. 2006. Transformaciones de la tierra, ensayos de historia ambiental. EUNED, San José, C.R, 264pp.

Wrigley, E.A. 1993. Cambio, continuidad y azar. Carácter de la Revolución industrial inglesa. Crítica, Barcelona, Spain. 180pp.

CHAPTER 5

The Economy of Forests

Sergio A. Molina-Murillo[1,*] and *Timothy M. Smith*[2]

"All wise forest management must have woodlands valued and endeavor to utilize them as much as possible, but in such a way that later generations will be able to derive at least as much benefit from them as the present generation claims for itself."

Georg Ludwig Hartig (1764–1837),
German agriculturist and writer on forestry

ABSTRACT

The economic value of forest goods and services is frequently ignored in market transactions and consequently undervalued in development strategies. Substantial scholar advancements in recent decades have brought a better understanding of forest functioning, its connection with other natural and human systems, and the development of alternative economic valuation techniques. Thus, our objective for this chapter is to discuss recent developments on the techniques for the economic valuation of forest goods and services and their monetary estimates. We based this chapter on the premise that protecting and properly managing forestland is paramount, and economic valuation is a viable and strong option for doing this. We frame our analysis on methodological aspects and also in the context of international

---
[1] Department of Environmental Sciences, National University of Costa Rica, Heredia 30101-Costa Rica; and Forest Resources Unit, University of Costa Rica, San Pedro de Montes de Oca, 11501-Costa Rica.
[2] NorthStar Initiative for Sustainable Enterprise, University of Minnesota, 1390 Eckles Ave., Saint Paul, MN 55108.
  E-mail: timsmith@umn.edu
* Corresponding author: sergiomolina@una.cr

socio-political agendas. We begin discussing the evolution of forests and forest valuation, and then detail the role of neoclassical economic theory for this valuation. Next, we describe economic frameworks that incorporate the valuation of ecosystem services. Finally, we provide some conclusions on the future of forest economics.

Introduction

Recent decades have witnessed a spectacular rise in both the techniques and concern for economic valuation of forest goods and services. While timber might be still the primary product from forests, benefits extracted from these ecosystems also include fuel, medicines, food, and other non-wood forest goods. In poor countries, forest products represent a large percentage of household income for many families (Vedeld et al. 2007; TEEB 2010). Forestland also supports human well-being through the continual provision of services such as water purification, carbon sequestration, and erosion control—to mention just a few. In the famous case of water provision for the city of New York, protecting the Catskill Mountains instead of installing expensive water filtration systems only made sense when the hidden costs and benefits were accounted for in the economic analysis (Appleton 2002).

Our objective for this chapter is to discuss recent developments in the techniques for the economic valuation of forest goods and services and their monetary estimates—the value itself. The economic value of forest goods and services is frequently ignored in market transactions and consequently undervalued in development strategies. Substantial scholarly advancements in recent decades have brought a better understanding of forest functioning, its connection with other natural and human systems, and the development of alternative economic valuation techniques. This highlights the need for using new techniques to guide practice and inform policy. Thus, this chapter is based on the premise that protecting and properly managing forestland is paramount, and economic valuation is a viable and strong option toward accomplishing this objective.

We frame our analysis on both methodological aspects, as well as the broader context of international socio-political valuation agendas. We begin by discussing the evolution of forests and forest valuation, and then detail the role of neoclassical economic theory for this valuation. Next, we describe economic frameworks that incorporate the valuation of ecosystem services. Finally, we provide some conclusions on the future of forest economics. Although we recognize there are additional ecosystem services besides those provided by forestland, in this chapter we use the terms forest services, ecosystems services, or environmental services interchangeably.

Our Changing View of Forest Economic Valuation

Humans have been closely connected with forests in multiple ways; nevertheless, our view towards them has substantially changed over time. As depicted in Fig. 1, this change occurred due to a better scientific understanding of its complexities, its prominent role in sustaining life on our planet, and our awareness of its gradual disappearance. A first distinct aspect within this changing view has to do with the use of forest products themselves. Throughout history, forests' main use, and consequently main value, has come from wood. Besides its most traditional use as fuel, wood is used to craft tools and instruments, and most commonly as a building material. Forests also provide a variety of other non-wood forest goods such as fibers, seeds, and medicines. Known as non-wood forest products (NWFPs), they have represented the main source of economic activity for many villages around the world.

Classical view		**Current view**
• Forests as sources of goods (e.g., wood, fibers, seeds)		• Forests as sources of goods and services (e.g., pollination)
• Ecosystems of local value		• Ecosystems of global value
• Abundant and replaceable		• Threatened and irreplaceable
• Market based		• Market & Non-Market based
• Mono-disciplinary << **Economic Valuation** >>		• Multi-disciplinary
• Single criterion		• Multiple criteria

Fig. 1. Changing elements of our economic view of forests.

However, in addition to forest goods, recently we have come to understand that forests provide a variety of services critical to maintain the proper functionality of ecosystems, and consequentially human well-being. Although conservationists have long called attention to forest preservation based on its ecological and intrinsic value, only recently have environmental services received significant attention by economists and policy makers, due to its connection with basic life supporting systems and the imminent threat to impair economic growth. Air purification, carbon sequestration, scenic beauty for tourism activities, water purification, flood control, biological services like pollination of agricultural plants, among many others, are examples of ecosystem services that are considered highly valuable today. In realizing this value, Costa Rica, for example, included several of these services in its 1997 forest law, in order to compensate forest owners for the additional services they provide to society (see Molina Murillo et al. 2014).

Spatial scale is a second aspect that influences our economic view of forests. Although forest goods have been exposed to international trade for centuries the connection between deforestation in a particular part of the world and its potential global consequences has just recently been recognized. With alarming rates of deforestation and the rapid disappearance of plant and animal species in tropical areas since the beginning of the twentieth century, the need for forest conservation and proper management became evident, in order to avoid a global catastrophe. Forest conservation has, thus, become a global effort supported by governments, non-governmental organizations, land-owners, research institutions, and even industry to a certain extent. This effort led to an increasing recognition that forests were connected to other global life supporting systems such as carbon, oxygen, and water cycles which in turn affect our global climate. Scientists have furthermore discovered that the forests' erosion-control capacity impacts river and coastal water quality, which directly correlates to the ocean's health and its ability to regulate climate and provide food.

The increasing threat of deforestation and disappearance of forest ecosystems is yet another element that significantly influences the way society economically values their protection and conservation. This threat is particularly severe for primary forests, ones which are largely irreplaceable. Although the increasing supply of timber from forest plantations may ease the logging pressure on primary forests, the two ecosystems are not fully comparable in their ecological functioning and the benefits they provide. Remnant primary forests are often subject to tight national and international regulations which encourage conservation and sustainable management. Following the scarcity principle of economics, this situation often creates greater value for banned species in black or unregulated markets, hampering the very objective of these regulations (Smith et al. 2013).

Based on scientific advancements, there are three other elements that have influenced the way society economically values forests. First, we have significantly progressed the valuation of forest using information that is not readily revealed by the market (e.g., people's willingness to protect the Amazon forest). These approaches are discussed in detail later in this chapter within valuation methodologies of ecosystem services. Second, the valuation of forest goods and services is now performed not only by pure economists, but by an array of researchers often done in coordination across a variety of multi-stakeholder groups (e.g., TEEB Initiative, The Natural Capital Project, The Ecosystem Service Partnership). Finally, studies valuing forest goods and services are more frequently conducted employing multiple criteria. Cost-benefit analysis based on the traditional economic view suggests that the benefits derived from a particular project or policy can compensate the costs inflicted, producing a net gain. However, as a complete economic valuation tool, cost-benefit analysis is problematic in most real-life situations

(Park and Gowdy 2013; Nunes et al. 2003) which require the transformation of all effects into one single monetary dimension/criterion (Munda 1996; de Marchi and Ravetz 2001). Thus, multi-criteria analysis is increasingly being employed as an appropriate tool for decision-making in complex policymaking situations (Pezzey 2004). Multi-criteria analysis is presented as a valuable method when assessing different existing alternatives, each one according to certain criteria (Getzner et al. 2005; Akgün et al. 2012).

These changes to the way society experiences forests and understands its benefits also influences the way economists estimate its value. In this regard, the classical economic valuation model employed to account for the welfare provided by nature is limited to adequately inform policies involving large-scale or long-term ecosystem changes such those provoked by climate change (Parker and Gowdy 2013; Gowdy 2005). From traditional economists to forest, natural, resource, and ecological economists, a key objective is to assess as fully and comprehensively as possible the value that forests provide to society. In the following sections we discuss in detail some of the techniques used to value forest products (both goods and services), some of the estimates of the value itself, and also some of the limitations most commonly encountered. As our intention is for this document to be easily understood by scholars and practitioners from many disciplines, we plan to use minimal economic and other jargon; however, we consider it important to introduce some theoretical economic ideas relevant to the understanding of forest valuation.

The Influence of Neoclassical Economics on Forest Valuation

Neoclassical economic theory centers on market allocation of products. It primarily focuses on how factors of production (e.g., labor, land, and capital) help to produce and distribute goods and services most efficiently. The parsimoniousness of the neoclassic economic model, however, has resulted in a series of gaps that severely hinder the sustainable use of natural capital-including forests. Efficiency is centered on the Pareto optimality concept, which means that no other allocation of resources would make at least one person better off without making someone else worse off. This concept, developed by Italian economist Vilfredo Pareto (1848–1923), relies on the invisible hand postulate by Adam Smith (Scottish Political Economist, 1723–1790), who theorized that the activity of individual actors pursuing their own best self-interest would also, as if guided by an invisible hand, achieve the best interest of the whole society with the least government intervention (*Laissez-faire*). Under this consideration, a free market system allows for efficient use of resources; however, this conceptualization is open to situations where market failures arise - particularly, in the case of forests, where benefits are a public goods or where the production/consumption of goods generate damages external to the transaction. For example, the public goods forests, e.g., biodiversity, carbon storage, and recreation, generate a market failure because their nonrival and nonexcludable characteristics prevent market incentives from achieving allocative efficiency. Similarly, pollution is a common example, where a gain by someone is achieved without offsetting (compensating) the loss of someone else (Harris 1996). Besides issues with efficiency itself, the fairness of the initial distribution, the source of the good, and the ultimate destination of product disposal (waste) are among other subjects not properly addressed by neoclassical theory.

Discount rates are yet another issue central to forest valuation. The neoclassical economic procedure of financial discounting reflects that, disregarding inflation, people value more immediate benefits over future benefits due to uncertainty in economic growth or impatient behavior. Discount rates are important in determining the rate at which renewable resources such as forests are used. The increase in the value of forests corresponds with the possibility of collecting gains by leaving the forest unharvested; that is, the untouched forest's value will appreciate if the price of its harvestable resources rises over time. However, if the discount rate is highly relative to its biological growth rate (stock growth), there will be an incentive to harvest now rather than wait, creating a potential scenario for forest extraction before optimal time.

In summary, when forests are valued only by their timber, other products are not incorporated into price calculations, falling outside the market. Since land owners are interested in profit maximization of timber, it is in their self-interest to pay the least amount possible for any factor of production. However, to the extent that low-cost production of timber creates costs and benefits external to the market transaction (e.g., pollution), depletes open-access public goods (e.g., deforestation in many parts of the world), or provides opportunities for free-riding (e.g., undetectable low priced illegal timber trade), the markets for harvested timber fail to increase societal welfare efficiently. Solutions to this problem have been addressed in the form of taxes (price regulation), cap-and-trade mechanisms (quantity regulation), property rights regulation, or through information access (forest certifications). These issues are particularly relevant to forests where costs of over-harvesting often cross property lines and where 80 percent of the world's forests are publically owned (FAO 2010). Acknowledging these limitations, scholars have attempted to estimate what the total value of forests might be and how it could be internalized into current markets.

Towards a Comprehensive Assessment of Forest Value

Forest goods and services provide multiple benefits to humankind. How much markets are willing to pay (WTP) in order to receive them depends on the value (satisfaction or utility) received or perceived from them. Although several classifications exist to segregate the total economic value (TEV) of forests, most of them begin by contrasting use and nonuse values. Use value refers to the utility or satisfaction that involves (directly or indirectly) a physical encounter with the forest. Examples of direct use value include timber or bushmeat; and examples of indirect use value include climate regulation or nutrient cycling. These uses can be consumptive and non-consumptive, that is, a wild animal inhabiting the forest could have non-consumptive value to tourists or it could have consumptive value to locals who eat it (Molina Murillo and Huson 2014). Nonuse value recognizes the value of certain benefits received from the forest even when not used. Two categories also occur here. Existence value (a.k.a. passive use value) refers to the satisfaction and WTP from the pure existence of a forest in a conserved state regardless of its current or planned use. One example might be the WTP for protecting the Amazon forests by European citizens even when they don't have the intention to ever visit South America. Bequest value, the other nonuse subcategory, refers to the value people receive when protecting forests for future generation to enjoy or use. Finally, there is another value, often included within both use and nonuse categories. This is known as option value and refers to a person's WTP for the option of using a forestland sometime in the future. In other words, option value is the future possible use in which the WTP is unrelated to present or projected use of the forest.

As shown, the value of forests is usually categorized according to the way satisfaction or utility is generated. In practice, however, limited evidence for some of these values exists, and studies have concentrated on the valuation of goods given its direct use and access to information.

Economic Valuation of Forest Goods

We start this section by discussing the methodologies used in the valuation of forest goods. Then, we present recent data on the monetary value of these goods on a global scale.

Methodological Approaches to the Valuation of Forest Goods

The valuation of forest goods is generally accomplished with information revealed by the market. For value calculation, a full series of costs (e.g., land, management, harvesting, transportation, processing, overhead, and others) are required in addition to possible streams of revenue. To explain the process of goods valuation, we turn to the valuation of timber, one of the most commonly used and traded forest goods.

In valuating timber production, general economic principles apply; although few peculiarities have been described by Robinson (1972). He argues that long harvesting periods, time and consequentially interest charges, are all important variables in forest management decisions. Since a promised dollar in the future is valued less than a dollar in the hand today, once someone invests in forest production this person surrenders the use of that money for some 20 to 50 years. The timber growth value in such case has to be higher than the opportunity costs, namely, the value of the best alternative—such as cattle ranching. This assessment is influenced by the individual or societal time preference, that is, the rate at which future income is discounted to make it comparable to present income. Another peculiarity of timber production pointed out by Robinson (1972) is that the forest is simultaneously the final product and the factory which produces the product. This has implications for protection cost because there is a high ratio of inventory (growing stock) to annual production (growth). Finally, as we have mentioned, forests themselves are not just producers of timber, but they produce multiple other goods and services, and are interlinked very closely with multiple ecosystems and stakeholders. This creates a vast array of challenges for timberland valuation, often not found in other business models.

Economic analysis centers on finding optimal rotation cycles for timber production, that is, the optimal harvest age of a forest. Inspired by the ideas of German forester Martin Faustmann (1849), many important economists such as Ohlin (1921), Fisher (1930), and Samuelson (1976) have evaluated this issue. Samuelson, for example, suggested that Faustmann found the expression for the present value of a perpetual forest to be a function of rotation age and not the optimal rotation age. The Faustmann Pressler-Ohlin theorem or FPO, elaborated by Johansson and Lofgren (1985), proposes that "a forest stand shall be harvested when the rate of change of its value with respect to time is equal to the interest on the value of the stand plus interest on the value of the forest land." Knowing that tree growth follows a logistic function (growth first increases and then decreases), the decision of when to harvest is done at the margin, where the forest will remain standing until the marginal cost equals the marginal benefit. In a simplified way, a marginal benefit occurs if the tree will grow in volume and be worth more the following year. A marginal cost, on the other hand, is the best alternative investment. Thus, as further described by Anderson (2006, p. 307), the optimal year for harvest is the first year in which the tree growth rate slows until it is equal to the annual rate of the return from the best alternative investment.

Net benefit analysis (a.k.a. benefit-costs analysis) is a simple way to evaluate timber investments, and is done by summing all costs and benefits incurred during the whole investment period. All future values are converted into present values using market interest rates, an exercise known as discounting. To make calculations more realistic, marginal costs also include the cost of planting, harvesting, the value of the land, insurance cost, and others. Similarly, the marginal benefit of waiting another year will also include a series of non-timber values such as recreation, climate regulation, and nutrient cycling. Harvesting policy should consider maximizing the social welfare by accounting also for citizens' values of these multiple amenities. Back in 1976, Hartman presented the evaluation of the optimal age of harvest while considering other forests products beyond timber. Considering these methodological improvements and their limitations, let us now examine the economic value estimates of forest goods.

Value Estimations of Forests Goods

Timber production is important to human well-being and economic growth across the globe. In 2010 it was estimated that about 31 percent of the world's total land area is covered by forest, corresponding to over 4 billion hectares. Unfortunately, the net change in forest area in the period between 2000 and 2010 was estimated at –5.2 million hectares per year. Particularly in South America and Africa, reductions in forest cover are much greater, with losses of 8.3 million hectares per year during the period between 1990 and 2000 (FAO 2011). Unless the economic value of conservation exceeds that of deforestation, this trend will be difficult to deter.

The cash value of global forest goods extraction totaled US$150 billion in 2011 (FAO 2015). In 2005 the value was US$122 (FAO 2010), and it was distributed almost evenly throughout Europe, North and Central America, and Asia (about $35 billion each) with the remaining amount distributed across South America ($6.9 billion), Africa ($4.8 billion), and Oceania ($3.1 billion). From the total value reported, industrial roundwood accounted for 71 percent, followed by non-wood forest products (NWFPs) at 15 percent, and woodfuel at 14 percent. Values of industrial roundwood varied between $30–70 per cubic meter with an average of $51. Woodfuel value, on the other hand, was estimated at $18 per cubic meter. Of those NWFPs reported, plants accounted for the majority (84 percent) including fruits, mushrooms, nuts, and berries. Honey, beeswax, and bushmeat are among animal goods used. An example of a commercial NWFP is goji, a herbal product that generated revenues of more than US$170 million in 2008 in the United States alone (Nutrition Business Journal 2010). The extraction of genetic material is another important NWFP (Pearce 2001). Based on Swanson's (1997) assessment of agribusinesses, Pearce calculated that the industrialized agricultural system requires a germplasm renewal from biodiversity every 12 years. This highlights the economic value of biodiversity as an insurance against the failure of current crop genetic stock.

Forest goods are easier to value in comparison to forest services, but still their value is often poorly estimated (Dasgupta and Duraiappah 2012). Cash sales of forest products are a meager indicator of the total human use of forests. Only 20 percent of forest income comes from cash sales of forest products, while 80 percent of that income includes products that never enter the market (FAO 2011). In Mexico, for example, woodfuel extracted for household consumption was found to be 17 times higher than that reported by official statistics (Gonzalez Martinez 2007). This is important considering that some 80–96 percent of woodfuel's consumers collect their own material (FAO 1996), and woodfuel is central for low-income families in poor countries.

Illegal activities and black markets are an additional limitation to properly valuing forest goods. Researchers from the World Bank (2011) estimated that annually we lose between US$10–15 billion from illegal cutting of forest world-wide, an activity which produces substantial damages to the land, a lowered income from sustainably managed forests, job losses, and uncollected taxes and royalties. Furthermore, if we consider NWFPs, the valuation is even more unreliable. Because available statistics often ignore non-commercial products, the estimates vary notably across regions, and the monetary value itself depends on the composition of tree species and the type of methodology employed in their valuation.

Economic Valuation of Forest Services

Ecosystem services are nature's contributions to societal well-being and which are perceived directly or indirectly as benefits (MEA 2005). Authors such as Barbier (2011) and de Groot et al. (2010) pointout that although valuing ecosystem services seems straightforward, in practice it is difficult. We start by reviewing key methodological aspects pertinent to the valuation of forest services, and then, present the monetary value of a series of services as examples.

Methodological Approaches to the Valuation of Forest Services

A critique to forest valuation argues that when it comes to formal benefit-cost analysis, marginal value is what is required, rather than aggregated global values (Pearce 2001). Furthermore, from a philosophical and ecological perspective, the value of forest goods and services is often intrinsic, making its valuation unrealistic. However, this valuation provides powerful arguments for forest conservation (e.g., Groot et al. 2012; Chee 2004; Balmford et al. 2002). In their study, Gómez-Baggethun et al. (2010) conclude that economic valuation and payment schemes have indeed contributed in attracting political support for conservation, but they also caution on the commoditization of forest services. They argue that this trend is partly the result of an economic conceptualization of nature's benefits from use values to exchange values. Others argue that economic approaches are useful in designing incentive schemes to make conservation policy both effective and efficient (Polasky et al. 2005; Costanza et al. 1997). For a discussion on risks and opportunities of economic valuation techniques for the valuation of ecosystem services please review Spangenberg and Settele (2010).

Although we agree that economic methods are useful to improve decision-making for resource use and conservation, several methodological issues are still unresolved. First, the classification of services is not yet clearly defined, causing problems for decision-making. Wallace (2007, 2008) argues for example, that some classifications are redundant because processes (means) for achieving services and the services themselves (ends) are included within the same classification category. Second, the valuation of forest services is often empirically tested in a simple yet incomplete manner (Barbier 2011). This usually occurs due to logistics in the field or in order to keep economic models parsimonious. However, the interaction of services with one another in nature is complex, non-linear, and involves multiple dimensions. Consider a forest providing climate regulation while at the same time supplying appropriate conditions for nutrient cycling or storm prevention. A forest provides numerous and overlapping services over a changing timeline. Turner et al. (2003) argue that the production function of nature is so complex that it is not possible to measure it in any reliable manner. A final methodological limitation on the valuation of forest services is the valuation of socio-cultural benefits (Carpenter et al. 2009). Just as we appreciate biophysical goods such as timber or water, nowadays biogeochemical processes such as carbon fixation are also recognized as socio-culturally valuable. Measured in cubic meters, liters per second, or tons per hectare, the economic valuation of these services is often related to a stock unit. However, straight-forward assessments are not yet solidly established for socio-cultural services produced by forests. Is a single tree worth more for a painter than a forest stand? Levels of aesthetic enjoyment or recreational and spiritual use are not directly connected to levels of forest stock. While these services might seem small when calculated as value per hectare, they are in fact substantial when considered as a percentage of total household income (e.g., Bahuguna 2000; Schaafsma et al. 2012; Mamo et al. 2007).

As a result of this complexity, multi-dimensionality, incomplete characterization, and difficulty of measurement, scholars have developed over the years a variety of economic methods to assess the value of forest services, which we summarized in Fig. 2 (e.g., Freeman 2003; Mendelsohn and Olmstead 2009). Although valuation methods of other nature exist (see EPA 2009, p. 42), we focus our attention here on the economic ones most commonly used in the valuation of forest services. These approaches are initially categorized based on the source of the information. If the data comes directly or indirectly from the market, they are known as revealed-preference methods; however, if the information is not found in the market but instead revealed by people when asked, they are known as stated-preference methods.

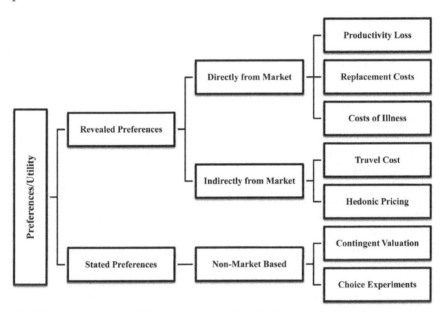

Fig. 2. Most commonly used forest/natural capital valuation methods.

Among the most common revealed-preference methods where information comes directly from the market are: productivity loss, replacement-costs, and cost-of-illness. Productivity loss examines the effects of the output of an economic activity, measured in terms of costs and profitability, which may be caused by a change in the quality/quantity of a forest product like water. The replacement-cost (RC) method states that if the environment has already been damaged, in order to restore it to its original state one must spend money. Thus, restoring cost is used as a proxy to estimate the value of the services provided by a forestland—for example cleaning up a lake. Cost-of-illness involves the assessment of all medical costs, other out-of-pocket expenses, and any loss of earnings due to an increase in morbidity. For example, all the costs of a disease caused by a mosquito outbreak after natural predators were significantly diminished because of forest destruction. All these expenses are revealed from the market in some way. Note that cost-of-illness estimates should be interpreted with caution, as they are normally just a conservative estimate of the true cost (benefit) associated with alterations of the forest.

Travel cost models and hedonic pricing are among the revealed-preference methods where information comes indirectly from the market, also known as surrogate markets. The travel cost model is most frequently applied in valuing natural environments that people visit to appreciate (e.g., tourism in the Tropical Cloud Forest Reserve). The basic premise is that time and travel costs represent the price or willingness to pay to visit the site. The hedonic pricing approach compares the value of a piece of land with multiple factors such as proximity to amenities, including the provision of forest goods and services.

Revealed-preference methods can only estimate the use-value of the environment, and only if this value affects human behaviour in a measurable and interpretable manner; thus, alternative methods have been developed employing hypothetical or experimental "markets." Contingent valuation and choice experiments are the most common methods within this category. Since the early work of Davis (1963) on the valuation of outdoor recreation, contingent valuation has become the predominant approach incorporating passive use value (Stenger et al. 2009). It involves asking people their willingness to pay for certain environmental goods and services provided by a forest, or their willingness to accept certain changes to it (e.g., forest flooding to a new hydroelectric reservoir). Choice experiments, also known as conjoint analyses or attribute-based methods, derive consumer utilities randomly from a bundle of attributes, where price and forest goods and services are included in the bundle (e.g., Adamowicz et al. 1998; Meyerhoff et al. 2009). These methodologies assume that individuals will rationally choose among bundles in order to maximize their utility. Although these methodologies have suffered criticism for their potential bias and unreliability (e.g., Smith 1993), in the absence of nonuse values and following a series of guidelines (Bateman and Turner 1993; NOAA 1993), they may be important for deriving useful information (Venkatachalam 2004).

Value Estimations of Forests Services

The benefits provided by forest ecosystems outweigh their protection cost by significant margins (Balmford et al. 2002). Since the presentation of the Millennium Ecosystem Assessment report in 2005, we have come a long way toward understanding and assessing the benefits that forest ecosystems provide to sustain human life. One of the first global attempts to quantify the value of world ecosystem services comes from Costanza et al. (1997), who estimated it to be in the range of US$16–54 trillion per year, with an average of US$33 trillion per year in 1995 U.S. dollars (equivalent to US$46 trillion per year in 2007 U.S. dollars). An updated estimate in 2011 resulted in US$145 trillion per year, also in 2007 U.S. dollars (Costanza et al. 2014). Similarly, researchers at the World Bank (2011) estimated that in 2005 the total economic value of natural assets at US$44 trillion worldwide.

Another set of studies conducted by the UN initiative, The Economics of Ecosystems and Biodiversity (TEEB), began in 2007 and aims to assess the economic impact of the global loss of biodiversity in order to present a convincing economic case for conservation (Ring et al. 2010). Among some global assessments resulting from this initiative include the estimation of the total global economic value of insect pollination, at €153 billion (Gallai et al. 2009), and the estimation of conserving forests to avoid greenhouse gas emissions worth US$3.7 trillion in net present value terms (Eliasch 2009). Unfortunately, forests continue to be under severe treat. The UN United Nations Convention to Combat Desertification (2013) reports that the cost of land degradation, understood as the reduction or loss in the biological or economic productive capacity of the land, is approximately US$500 billion per year, impacting directly some 1.5 billion people.

In Table 1 we provide some examples of recent research on economic valuation across different forest services. Yet, we consider it important to clarify that the results should be treated as good approximations because of methodological limitations (Turner et al. 2009), because they are context- and scale-dependent, and because ultimately, preferences are may change, and even could be formed during the valuation process itself (Parker and Gowdy 2013). For those interested, Söderqvist and Soutukorva (2009) provides a thought-provoking discussion on assessing the quality of environmental valuation studies. For a comprehensive meta-analysis of recent estimates across multiple biomes and services we recommend the exploration of de Groot (2012) and for estimations conducted before the year 2000, you may explore Pearce (2001), Torras (2000), and Nunes et al. (2003). As shown in Table 1, the variation in the value could be substantial across services; however, all of these studies provide important tool applications, and in general, they provide support for the premise that the value of these services diminish with forest loss.

Table 1. Examples of recent economic valuation of forest services.

Forest Service	Valuation Method	Value (US$)	Reference	Observations
Water regulation and provision	Contingent Valuation	0.64/Ha/yr	Vargas (2004)	Economic valuation based on Payment for Ecosystems Services in Bolivia.
Climate regulation	Productivity loss based on cooling energy savings	12,232/Ha/yr	Chen (2006)	Based on the Guangzhou city of China with a nearby forest of 7,360 ha. Using an average 2006 exchange rate conversion (1 RMB ≅ 0.16 US$).
Pollination	Productivity loss	33/Ha/yr	Olschewski et al. (2010)	Value for coffee farming in tropical forest.
Storm prevention	Replacement cost	3,700 Ha	Sathirathai and Barbier (2001)	Mangroves as barriers for cyclones and storms in Thailand.
Flood protection	Replacement cost	24/Ha/yr	Yaron (2001)	Study conducted in Cameroon.
Pharmaceutical value	Productivity loss	~6/Ha	Croitoru (2007)	For Turkey, 2001 base year used.
Carbon sequestration	Productivity loss	12–27 /Ha/yr	Sohngen and Brown (2006)	Based on subsidies to maintain forestland in South Central United States.
Scenic beauty	Choice experiment	69.8–110.12/ household/yr	Juutinen et al. (2014)	Citizens' valuations of recreation-oriented management in state-owned commercial forests in Finland. Using an summer 2012 exchange rate conversion (1 EU ≅ 1.25 US$).

Category	Method	Value	Reference	Notes
Scenic beauty	Meta-analysis using Travel Cost Method	0.81–138 per trip	Zandersen and Tol (2009)	Countries across Europe. Using an average 2000 exchange rate conversion (1 Euro ≅ 1.23 US$).
Water quality	Meta-analysis based on studies that use Contingent Valuation and Choice Experiments	11–13 for nonusers to 32–37 for users, per one-unit change in water quality	Van Houtven et al. (2007)	Estimates based on log-linear function on the predicted mean WTP estimates for a one-unit change in water quality.
All goods and services-Forests as a whole	Contingent valuation	67.94/person/yr	Gürlük (2006)	Based on a Rural Development project in northwest Turkey.
	Combination of revealed and stated approaches	162.26/Ha/yr	Croitoru (2007)	Based on information from 17 Mediterranean countries. Using an average 2001 exchange rate conversion which was the base year in the study (1 Euro ≅ 1.22 US$).
	Choice experiments	19/Ha/yr	Mallawaarachchi et al. (2001)	Study conducted in North Queensland Australia. Using an average 2010 exchange rate conversion (1 AU$ ≅ 1.05 US$).

The Future of Forest Economics

Recent assessments suggest that the substantial value of forest goods and services is not included within their valuation (Sukhdev 2011; Costanza et al. 2014), which contributes to the widespread neglect and degradation of the very forests producing these goods and services. Given the complexity and magnitude of the impacts caused by their destruction, a strong case exists for their valuation and inclusion in business and national accounting systems. Fortunately, the future seems promising. For example, at the Rio +20 conference, the apparel company Puma presented the first-of-its-kind attempt to monetize the value of ecosystem services used across the company's entire supply chain. Known as the Environmental Profit & Loss Account, Puma estimated this cost at about €145 million in 2010. At the national level, the World Bank through the WAVES partnership, which stands for Wealth Accounting and the Valuation of Ecosystem Services, is aiming to include the value of natural resources into national accounting systems. However, accounting the wealth of forest ecosystems is insufficient if global market systems disproportionately distribute the benefits. Developed countries could support or compensate developing countries that incur net losses from conservation policies. Some mechanisms are international resource transfers such as debt-for-nature swaps, donations for conservation, official aid, or through pricing mechanisms and the elimination of perverse subsidies on agriculture and energy. This is paramount considering that for many fast-growing developing countries the pressure on forestland and forest resources continuously increases.

Substantial progress has been made also in terms of valuation methodologies; although, further refining is necessary. First, economic valuation of environmental services is significantly impaired by our understanding of their structure, function, and their production flow to people (Barbier 2011). Second, for certain goods and services, methodological generalization beyond specific scenarios is vague. This is particularly important considering that most of the value generated by standing forests tends to be in the form of public goods. Finally, when assessing the value of forests for future use and nonuse, discount rates must be carefully considered and managed. Since high income individuals and rich societies tend to discount at lower rates, given their diminishing utility of additional income, researchers and policy makers need to wisely evaluate each particular context in which actions will promote sustainable forest use. Pearce (2001), for example, suggests targeted low-cost credit to agricultural colonists as a mechanism to reduce discount rates.

With valuation methodologies further developing and thus proving an economic case for the sustainable use of forests, their use requires caution in order to avoid the commoditization of forest resources. In this regard, the sustainable use of forest ecosystems and our understanding of their contribution to human well-being must precede any efforts in this direction. The survival of the fittest is dependent on this shift of paradigm.

Acknowledgments

We are very grateful to Claire Fox, Research Associate at SFS-Costa Rica, and Samantha Steinbring, Communications Specialist, at the NorthStar Initiative, University of Minnesota, for revising and editing earlier versions of this chapter and providing helpful suggestions.

References

Adamowicz, W.L., P. Boxall, M. Williams and J. Louviere. 1998. State preference approaches for measuring passive use values: choice experiments and contingent valuation. Am. J. Agr. Econ. 80: 64–75.

Akgün, A.A., E. van Leeuwen and P. Nijkamp. 2012. A multi-actor multi-criteria scenario analysis of regional sustainable resource policy. Ecol. Econ. 78: 19–28.

Ambrey, C.L. and C.M. Fleming. 2011. Valuing scenic amenity using life satisfaction data. Ecol. Econ. 72: 106–115.

Anderson, D.A. 2006. Environmental Economics and Natural Resource Management. Pensive Press, Danville, Kentucky.

Appleton, A. 2002. How New York City Used an Ecosystem Services Strategy Carried out Through an Urban-Rural Partnership to Preserve the Pristine Quality of Its Drinking Water and Save Billions of Dollars and What Lessons It Teaches about Using Ecosystem Services. Paper presented at The Katoomba Conference. Tokyo, 13pp.

Bahuguna, V.K. 2000. Forests in the economy of the rural poor: an estimation of the dependency level. AMBIO 29: 126–129.

Balmford, A., A. Bruner, P. Cooper, R. Costanza, S. Farber, R.E. Green, M. Jenkins, P. Jefferiss, V. Jessamy, J. Madden, K. Munro, N. Myers, S. Naeem, J. Paavola, M. Rayment, S. Rosendo, J. Roughgarden, K. Rumper and R.K. Turner. 2002. Economic reasons for conserving wild nature. Science 297: 950–953.

Barbier, E.B. 2011. Pricing nature. Annu. Rev. Resour. Econ. 3: 337–353.

Bateman, I.J. and R.K. Turner. 1993. Valuation of environment, methods and techniques: the contingent valuation method. pp. 120–191. In: R.K. Turner (ed.). Sustainable Environmental Economics and Management: Principles and Practice. Belhaven Press, London.

Carpenter, S.R., H.A. Mooney, J. Agard, D. Capistrano, R.S. DeFries, S. Díaz, T. Dietz, A.K. Duraiappah, A. Oteng-Yeboah, H.M. Pereira, C. Perrings, W.V. Reid, J. Sarukhan, R.J. Scholes and A. Whyte. 2009. Science for managing ecosystem services: beyond the millennium ecosystem assessment. P. Natl. Acad. Sci. USA 106: 1305–1312.

Chee, Y.E. 2004. An ecological perspective on the valuation of ecosystem services. Biol. Conserv. 120: 549–565.

Chen, W.Y. 2006. Assessing the Services and Value of Green Spaces in Urban Ecosystem: A Case of Guangzhou City. Ph.D. Thesis, University of Hong Kong, Hong Kong.

Costanza, R., R. d' Arge, R. de Groot, S. Farber, M. Grasso, B. Hannon, K. Limburg, S. Naeem, R. O'Neill, J. Paruelo, R. Raskin, P. Sutton and M. van den Belt. 1997. The value of the world's ecosystem services and natural capital. Nature 387: 253–260.

Costanza, R., R. de Groot, P. Sutton, S. van der Ploeg, S. Anderson, I. Kubiszewski, S. Farber and R.K. Turner. 2014. Changes in the global value of ecosystem services. Global Environ. Chang. 26: 152–158.

Croitoru, L. 2007. How much are Mediterranean forests worth? Forest Policy Econ. 9: 536–545.

Dasgupta, P. and A. Duraiappah. 2012. Well-being and wealth. pp. 13–26. *In*: UNU-IHDP and UNEP (eds.). Inclusive Wealth Report 2012: Measuring Progress Toward Sustainability. Cambridge University Press, Cambridge.

Davis, R.K. 1963. Recreation planning as an economic problem. Nat. Resour. J. 3: 239–249.

de Groot, R., R. Alkemade, L. Braat, L. Hein and L. Willemen. 2010. Challenges in integrating the concept of ecosystem services and values in landscape planning, management and decision making. Ecol. Complex 7: 260–272.

de Groot, R., L. Brander, S. van der Ploeg, R. Costanza, F. Bernard, L. Braat, M. Christie, N. Crossman, A. Ghermandi, L. Hein, S. Hussain, P. Kumar, A. McVittie, R. Portela, L.C. Rodriguez, P. ten Brink and P. van Beukering. 2012. Global estimates of the value of ecosystems and their services in monetary units. Ecosystem Services 1: 50–61.

de Marchi, B. and J. Ravetz. 2001. Participatory approaches to environmental policy. Environmental Valuation in Europe, Policy Research Brief #10. Cambridge Research for the Environment, Italy.

Eliasch, J. 2009. Climate Change: Financing Global Forests: The Eliasch Review. Earthscan, United Kingdom.

[EPA] Environmental Protection Agency. 2009. Valuing the Protection of Ecological Systems and Services. A Report of the EPA Science Advisory Board. EPA-SAB-09-012, EPA, Washington, D.C.

[FAO] Food and Agricultural Organization of the United Nations. 1996. Memoria—Reunión regional sobre generación de electricidad a partir de biomasa. Informe México-Balance Energético Nacional, FAO-Forestal. Santiago, Chile.

[FAO] Food and Agricultural Organization of the United Nations. 2010. Global Forest Resources Assessment, 2010—Main Report. FAO Forestry Paper 163. Rome, Italy, 343pp.

[FAO] Food and Agricultural Organization of the United Nations. 2011. State of the World's Forests 2011. Rome, Italy, 179pp.

[FAO] Food and Agricultural Organization of the United Nations. 2015. Global Forest Resources Assessment 2015: How are the world's forests changing? Rome, Italy, 49pp.

Faustmann, M. 1849. On the Determination of the Value Which Forest Land and Immature Stands Possess for Forestry. English edition edited by M. Gane. Oxford Institute Paper 42 (1968).

Fisher, I. 1930. The Theory of Interest. Macmillan, New York.

Freeman, A.M. III. 2003. The Measurement of Environmental and Resource Values: Theory and Methods. Resources for the Future, Washington, DC.

Gallai, N., J.M. Salles, J. Settele and B.E. Vaissière. 2009. Economic valuation of the vulnerability of world agriculture confronted with pollinator decline. Ecol. Econ. 68: 810–821.

Getzner, M., C.I. Spash and S. Stagl. 2005. Alternatives for Environmental Valuation. Routledge, London.

Gómez-Baggethun, E., R. de Groot, P.L. Lomas and C. Montes. 2010. The history of ecosystem services in economic theory and practice: from early notions to markets and payment schemes. Ecol. Econ. 69: 1209–1218.

González Martínez, A.C. 2007. La extracción y consumo de biomasa en México (1970–2003). Integrando la leña en la contabilidad de flujos de materiales. Rev. Iberoam Econ. Ecol. 6: 1–16.

Gowdy, J. 2005. Toward a new welfare economics for sustainability. Ecol. Econ. 53: 211–222.

Gürlük, S. 2006. The estimation of ecosystem services' value in the region of Misi Rural Development Project: results from a contingent valuation survey. Forest Policy Econ. 9: 209–218.

Harris, M. 1996. Environmental Economics. Aust. Econ. Rev. 29: 449–465.

Hartman, R. 1976. The harvesting decision when a standing forest has value. Econ. Inq. 14: 52–58.

Johansson, P.O. and K.G. Lofgren. 1985. The Economics of Forestry and Natural Resources. Basil Blackwell, Oxford, United Kingdom.

Juutinen, A., A.K. Kosenius and V. Ovaskainen. 2014. Estimating the benefits of recreation-oriented management in state-owned commercial forests in Finland: a choice experiment. J. For. Econ. 20: 396–412.

Mallawaarachchi, T., R.K. Blamey, M.D. Morrison, A.K.L. Johnson and J.W. Bennett. 2001. Community values for environmental protection in a cane farming catchment in Northern Australia: a choice modelling study. J. Environ. Manage. 62: 301–316.

Mamo, G., E. Sjaastad and P. Vedeld. 2007. Economic dependence on forest resources: a case from Dendi District, Ethiopia. Forest Policy Econ. 9: 916–927.

[MEA] Millennium Ecosystem Assessment. 2005. Ecosystems and Human Well-being: General Synthesis. Island Press, Washington, DC.

Mendelsohn, R. and S. Olmstead. 2009. The economic valuation of environmental amenities and disamenities: methods and applications. Annu. Rev. Env. Resour. 34: 325–347.

Meyerhoff, J., U. Liebe and V. Hartje. 2009. Benefits of biodiversity enhancement of nature-oriented silviculture: evidence from two choice experiments in Germany. J. Forest Econ. 15: 37–58.

Molina Murillo, S.A. and K. Huson. 2014. Poaching, rural communities and tourism development: a case study in Costa Rica. Inter. J. Develop. and Sust. 3: 1287–1302.

Molina Murillo, S.A., J.P. Perez Castillo and M.E. Herrera Ugalde. 2014. Assessment of environmental payments on indigenous territories: the case of Cabecar-Talamanca, Costa Rica. Ecosystems Serv. 8: 35–43.

Munda, G. 1996. Cost-benefit analysis in integrated environmental assessment: some methodological issues. Ecol. Econ. 19: 157–168.

[NOAA] National Oceanic and Atmospheric Administration. 1993. Report of the NOAA Panel on contingent valuation. Federal Register 58: 4602–4614.

Nunes, P.A.L.D., J.C.J.M. van den Bergh and P. Nijkamp. 2003. The Ecological Economics of Biodiversity. Methods and Policy Applications. Edward Elgar, Cheltenham, United Kingdom.

Nutrition Business Journal. 2010. '09 Sales Growth Sputters in Every Nutrition Category as Economy Takes its Toll. June 1, 2010. pp. 1.

Ohlin, B. 1921. Till Fragan om Skogarnas Omloppstid (On the Question of the Rotation Period of the Forests). Ekonomisk Tidskrift 12: 89–113.

Olschewski, R., A.M. Klein and T. Tscharntke. 2010. Economic trade-offs between carbon sequestration, timber production, and crop pollination in tropical forested landscapes. Ecol. Complex 7: 314–319.

Parker, S. and J. Gowdy. 2013. What have economists learned about valuing nature? A review essay. Ecosystem Services 3: e1–e10.

Pearce, D.W. 2001. The economic value of forest ecosystems. Ecosys. Health 7: 284–296.

Pezzey, J.C.V. 2004. Sustainability policy and environmental policy. Scand. J. Econ. 106: 339–359.

Pereira-Goncalves, M., M. Panjer, T.S. Greenberg and W.B. Magrath. 2012. Justice for Forest: Improving Criminal Justice Efforts to Combat Illegal Logging. The World Bank, Washington, DC.

Polasky, S., C. Costello and A. Solow. 2005. The economics of biodiversity. pp. 1517–1560. *In*: K.G. Mäler and J.R. Vincent (eds.). Handbook of Environmental Economics, Volume 3: Economywide and International Environmental Issues. Elsevier, North Holland.

Ring, I., B. Hansjürgens, T. Elmqvist, H. Wittmer and P. Sukhdev. 2010. Challenges in framing the economics of ecosystems and biodiversity: the TEEB initiative. Curr. Opin. Environ. Sustain. 2: 15–26.

Robinson, G.G. 1972. Forest Resource Economics. The Ronald Press Co. New York, New York.

Samuelson, P.A. 1976. Economics of forestry in an evolving society. Econ. Inq. 14: 466–492.

Sathirathai, S. and E.B. Barbier. 2001. Valuing mangrove conservation in southern Thailand. Contemp. Econ. Policy 19: 109–122.

Schaafsma, M., S. Morse-Jones, P. Posen, R.D. Swetnam, A. Balmford, I.J. Bateman, N.D. Burgess, S.A.O. Chamshamae, B. Fishera, R.E. Greenb, A.S. Hepelwah, A. Hernández-

Sirventi, G.C. Kajembee, K. Kulindwah, J.F. Lundk, L. Mbwambol, H. Meilbyk, Y.M. Ngagae, I. Theiladek, T. Treuek, V.G. Vyamanae and R.K. Turner. 2012. Towards transferable functions for extraction of non-timber forest products: a case study on charcoal production in Tanzania. Ecol. Econ. 80: 48–62.

Smith, K.V. 1993. Nonmarket valuation of environmental resources: an interpretative appraisal. Land Econ. 69: 1–26.

Smith, T.M., S.A. Molina Murillo and B.M. Anderson. 2013. Implementing sustainability in the forestry sector: toward the convergence of public and private forest policy. pp. 237–260. *In*: E. Hansen, R. Panwar and R. Vlosky (eds.). The Global Forest Sector: Changes, Practices, and Prospects. Taylor and Francis, Florida.

Söderqvist, T. and A. Soutukorva. 2009. On how to assess the quality of environmental valuation studies. J. Forest Econ. 15: 15–36.

Sohngen, B. and S. Brown. 2006. The influence of conversion of forest types on carbon sequestration and other ecosystem services in the South Central United States. Ecol. Econ. 57: 698–708.

Spangenberg, J. and J. Settele. 2010. Precisely incorrect? Monetizing the value of ecosystem services. Ecol. Complexity 7: 327–337.

Stenger, A., P. Harou and S. Navrud. 2009. Valuing environmental goods and services derived from the forests. J. Forest Econ. 15: 1–14.

Sukhdev, P. 2011. Putting a price on nature: the economics of ecosystems and biodiversity. Solutions 1: 34–43.

Swanson, T. 1997. Global Action for Biodiversity. Earth-Scan, London, United Kingdom.

[TEEB] The Economics of Ecosystems and Biodiversity. 2010. The Economics of Ecosystems and Biodiversity: Mainstreaming the Economics of Nature: A Synthesis of the Approach, Conclusions and Recommendations of TEEB. Malta, 36pp.

Torras, M. 2000. The total economic value of Amazonian deforestation, 1978–1993. Ecol. Econ. 33: 283–297.

Turner, R.K., J. Paavola, P. Cooper, S. Farber, V. Jessamy and S. Georgiou. 2003. Valuing nature: lessons learned and future research directions. Ecol. Econ. 46: 493–510.

Turner, R.K., S. Morse-Jones and B. Fisher. 2010. Ecosystem valuation, a sequential decision support system and quality assessment issues. Ann. NY Acad. Sci. 1185: 79–101.

[UNCCD] United Nations Convention to Combat Desertification. 2013. The Economics of Desertification, Land Degradation and Drought: Methodologies and Analysis for Decision-Making. Background Document. UNCCD 2nd Scientific Conference.

Van Houtven, G., J. Powers and S.K. Pattanayak. 2007. Valuing water quality improvements in the United States using meta-analysis: is the glass half-full or half-empty for national policy analysis? Resour. Energy Econ. 29: 206–228.

Vargas, M.T. 2004. Evaluating the economic basis for payments for watershed services around Amboró National Park, Bolivia. Master Thesis, Yale University, New Haven, Connecticut.

Vedeld, P., A. Angelsen, J. Bojo, E. Sjaastad and G.K. Berg. 2007. Forest environmental incomes and the rural poor. Forest Policy Econ. 9: 869–879.

Venkatachalam, L. 2004. The contingent valuation method: a review. Environ. Impact Assess. 24: 89–124.

Wallace, K.J. 2007. Classification of ecosystem services: problems and solutions. Biol. Conserv. 139: 235–246.

Wallace, K.J. 2008. Ecosystem services: multiple classifications or confusion? Biol. Conserv. 141: 353–354.

World Bank. 2011. The Changing Wealth of Nations: Measuring Sustainable Development in the New Millennium. Washington, D.C.

Yaron, G. 2001. Forest, plantation crops or small-scale agriculture? An economic analysis of alternative land use options in the Mount Cameroun Area. J. Environ. Plann. Man. 44: 85–108.

Zandersen, M. and R.S.J. Tol. 2009. A meta-analysis of forest recreation values in Europe. J. Forest Econ. 15: 109–130.

CHAPTER 6

Silviculture and Forest Management

Adrián A. Monge Monge

ABSTRACT

The evolution of silviculture has been a long learning process, not only technically, but also at a more philosophical level. In the west, much has changed since Hans Carl von Carlowitz published in 1730 his management manual to sustain wood supply for the local mining industry. After several attempts of domestication, homogenization and economic optimization of forest stands—some more successful than others—we are coming to the realization that the 'ideal' management of forest stands requires a more or less permanent forest cover and certain level of 'naturalness' as societies have come to expect much more than just a sustain supply of timber from forests. The following pages summarize the basic ecological principles governing forest silviculture and present a number of management examples from several regions to illustrate how forest management has been moving from a conversion approach towards a lower impact management. Nonetheless, in order to succeed any management regime should consider the needs and expectations of those closer or most dependent on the forest for their wellbeing; conversion, homogenization, rehabilitation and conservation of forest stands are all part of a dynamic forest landscape where

Department of Forest Sciences, University of Helsinki, P.O. Box 27, FI-00014 Helsinki, Finland.
 E-mail: adrian.mongemonge@helsinki.fi

trade-offs are unavoidable. The technical knowledge to managed forest stands in a continuous manner is available despite our incomplete ecological knowledge at certain levels; however, as with many other issues affecting the forestry sector, forest management is mostly limited by administrative and economic shortcomings than by the lack of detailed ecological knowledge. Conflicted legislation, complicated and costly bureaucratic requirements, and unreasonable expectations of short-term financial returns are undermining the long-term productive potential of many forest areas.

Introduction

Early references to regulations on forest use can be found during the expansion of the Roman Empire when wood became the key raw material that sustained the metallurgical, infrastructural, and army industries (particularly the navy), in a time when large forested areas were also cleared around cities for agricultural expansion (Grebner et al. 2013). These developments generated wood shortages in several parts of Europe to the point of the implementation of regulations controlling the cutting and planting of trees, as well as wood consumption reduction practices (Whited et al. 2005). Later examples of forest legislation can be found in the Venetian Republic in the sixteenth century in order to maintain timber supply for the shipping industry in the Mediterranean (Agnoletti et al. 2009). However, the idea of forest management as we know it today was introduced in Europe around 1730 with the publication of the book *Sylvicultura Oeconomica* by Hans Carl von Carlowitz that intended to sustain the supply of wood for the mining industry (Agnoletti et al. 2009). Outside Europe, examples of regulations on forest utilization can be found in China as early as the Qin and Han Dynasties (BC 221–AD 220) including subsidies for planting trees. These regulations changed and evolved until the introduction of more formal legislation during the Ming Dynasty (AD 1368–1644) and afterwards (Zhang 2001).

In the tropics, forest management was introduced during colonial times, after European experiences, in order to generate revenues from the colonies. In recent times, the maintenance and management of tropical forest resources has been a long-term topic of discussion among foresters, ecologists, conservationists, and social scientists (Unesco 1978; National Research Council 1980; Budowski 1984; Mergen 1987; Srivastava et al. 1982; Hallsworth 1982; Nabuurs et al. 2007). However, the general picture has not changed much since Gómez-Pompa and Burley (1991) argued that only a small fraction of tropical forests are under sustained regimes of wood production and the majority of forest areas are still harvested without real strategies for regeneration or replacement, despite intensive research in the last 50 years.

In this chapter, a revision of some of the basic ecological principles governing forest management are presented in the light of management systems that have been applied in different parts of the world. The chapter starts by setting a reference framework, followed by the introduction of ecological principles as well as general concepts within the topic of silviculture. The last section is a presentation of selective management systems based on a general classification.

For reference proposes, the objectives of forest management as defined herein are to enable the sustainable production of both timber and non-timber forest products (NTFP) and to ensure the supply of ecosystem services provided by forested areas (restoration and maintenance of natural processes). This chapter deals with silviculture and forest management under the classification suggested by Bawa and Krugman (1991). As part of this classification (Fig. 1), only intervention under the classes 1-2 and 3 will be considered since they enhance the supply of goods or services from the forest areas under management.

Over the years, forest management has been moving from a conversion approach to a lower impact management in which even secondary forests are managed to resemble mature forests and not homogenous forest stands (e.g., Central America). More recently, the inclusion of forest management as part of the mitigation activities under the United Nations Framework Convention on Climate Change (UNFCCC), has mobilized resources for the improvement of the forest sector in several countries in order to reduce

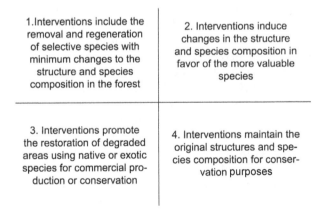

Fig. 1. Types of interventions based on the objectives of forest management, modified from Bawa and Krugman (1991).

emissions from land use change and enhance carbon stocks in managed forest areas.

Under the views of Webb (1982) and Kareiva et al. (2007), the non-pristine nature of most forests can be considered the pattern, since these have been utilized for centuries or decades by endemic tribes or other human groups. This will render most of what we would call natural, pristine, primary, or mature forests as an intrinsic mosaic of patches recovering from either natural or man-made disturbances. These patches are very diverse in terms of size and level of disturbance.

Gómez-Pompa and Burley (1991) mention three prerequisites any country would need to fulfill in order to ensure long-term productivity from forested areas. The first of these prerequisites is a forestry law or policy clearly stating the long-term commitment for the sustainable management of public and private forests. The second established norm would be a set of regulations or guidelines for forest management as well as reasonable procedures that reflect the long-term policy commitment. The third prerequisite is the allocation of sufficient human and financial resources for monitoring and implementation of both national policies and guidelines. Over the years more and more countries have achieved (at least on paper and at different levels of success) the first two requirements. However, in most cases, sufficient resources have not been allocated to corroborate whether policy and regulatory frameworks are adequate.

Ecological Principles

The key ecological principles governing forest management are the ones related to gap dynamics (Watt 1947; Whitmore 1984; Lamprecht 1989; Oldeman 1990; Matthews 1991), which further expand the idea that natural forest ecosystems are a dynamic mosaic of gaps at different stages of development or recovery. Several descriptions of the sequence of phases involved in gap dynamics have been presented over the years (Whitmore 1984; Lamprecht 1989; Oldeman 1990). In general these descriptions include an initial opening of the gap and the subsequent establishment of species favored by the newly created microclimatic and physical conditions and the closing of the new/young canopy dominated by long-survival early comers and the parallel disappearance of the short lived or light demanding species. Eventually the gap would resemble the surrounding forest as the vertical growth reaches its potential and horizontal growth dominates competition among canopy species and finally the long term survivor species start to become old and small to medium gaps start to appear to eventually connect to the opening of a new large gap.

The phases described above are of course an oversimplification of the dynamic changes affecting any particular forest stand. For example, the edges of a gap are not static during the recovery process and fresh gaps (large or small) could appear next to the old, thus affecting the development of the original gap by favoring some species or delaying the establishment of others. It is also well recognized that the size of the gap influences the mixture of species filling the open area or taking advantage of the newly available resources. Whitmore (1984 citing Kramer 1933), recalls an experiment in a lower mountain forest (West Java, Indonesia) with man-made gaps of 0.1, 0.2, and 0.3 ha, where seedling composition in the small gaps resembled the tree composition of the canopy, while in larger gaps a new composition of seedlings (with many newcomers) was found. Swaine and Whitmore (1988) classified these species as pioneers or those whose seedlings will appear only after the opening of gaps; and non-pioneers (climax species) or those whose seedlings are growing below the canopy. These authors also suggested a continuous variation within each group in a manner that a further subdivision is also possible.

At this point, it is not hard to imagine that light availability plays an important role in the separation between pioneer and non-pioneer species. This is the reason why a number of forest ecologists had used light dependency (or shade tolerance) to classify tree species. Authors such as Baur (1964), Mayer (1980), and Oldeman (1990) shown in Table 1 demonstrate how species can been classified based on their light requirements (known as temperaments) along their life cycles. In this sense, it is not rare to read terms such as strict heliophilous, semi heliophilous, and long-lived heliophilous in relation to light demanding trees, and terms such as semi-sciophilous, and true sciophilous in relation to shade tolerant species. Furthermore, shade-tolerant species are sometimes subdivided into canopy species and understory species to separate potentially commercial species from smaller trees. Similarly, terms such as long-lived heliophilous or long-lived pioneers are used to indicate commercial species (Budowski 1970 also refers to them as late-secondary species).

Shugart (1984), working on the modeling of gap dynamics, further complemented the previous classifications by combining the potential hold by different species to generate significant gaps as well as their dependency on gaps for establishment. Generally speaking, commercial trees for logging are very likely gap-makers, but some of them do not need a gap for regeneration. In this way, four new roles can be identified as shown in Fig. 2.

It is worth mentioning that more detailed classifications have been made over the years. Hallé et al. (1978) used development trends in the architecture of trees and tried to correlate them to some of the temperaments. However, the classification is hard to put in practice without sufficient experience,

Table 1. Classification of forests tree species based on their light requirements or shade tolerance.

Baur (1964)	Mayer (1980)	Oldeman (1990)	Generalities
Secondary species	Light demanding (Lichtbaumarten)	Light demanding	- Require light for most of their life cycles - Established and grow-up in large gaps - Produce abundant seeds at an early age - Good dispersability of seeds - Seeds maintain viability for long periods
Gap-opportunist	Half-shade tolerant (Halbschattbaumarten)	Semi tolerant	- Can survive some shade during germination and establishment - If light conditions do not improve after some time, they die - High seed production, either frequent or periodical (maintaining a stable population of seedlings)
Truly-tolerant species	Shade-tolerant (Schattbaumarten)	Shade-tolerant	- Can germinate and survive for many years under heavy shade - Respond to improving light conditions by increasing growth - Can reach up to medium and canopy size trees - Not frequent or high seed production
		Late-tolerant	- Require light for germination and establishment - Can survive in the shade for a very long time

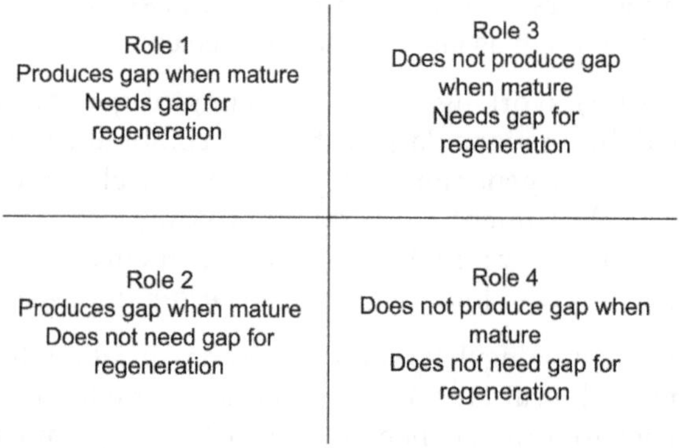

Role 1 Produces gap when mature Needs gap for regeneration	**Role 3** Does not produce gap when mature Needs gap for regeneration
Role 2 Produces gap when mature Does not need gap for regeneration	**Role 4** Does not produce gap when mature Does not need gap for regeneration

Fig. 2. Classification of trees based on their capacity to produce large gaps or being dependent on gaps for stabilizing regeneration, modified from Shugart (1984).

as the crown shape is easily affected by other interacting elements in the forest. Oldeman and van Dijk (1991) suggested an alternative classification by combining the vegetative life story and reproductive strategies of tree species; between the extremes hard gamblers or light demanding species and the hard strugglers or shade tolerant. There are four intermediate groups and they allow for changes in light dependency or shade tolerance as trees mature and crown and leaf characteristics are considered for this classification.

Another ecological consideration in terms of successional modeling is the set of three mechanisms suggested by Connell and Slatyer (1977) where early colonizers of a gap, or even resident species, either facilitate, tolerate, or inhibit the establishment of late successional species (Fig. 3). These mechanisms are closely related to the concepts of autogenic succession (Egler 1954) and allogenic succession (Spurr and Barnes 1980). In these cases, succession relays on the recruitment of seedlings from trees already developing under the original canopy (autogenic), or on the recruitment of seedlings from species arriving only after the creation of a gap (allogenic) and these new arrivals overtake the initial floristic composition and change it over time. In practice, it is likely that autogenic processes dominate in small disturbances while allogenic processes are more relevant as the size or extent of the disturbance increases.

As mentioned before, the size of the gap, also referred as eco-unit, has a significant relevance in determining the variability of the species

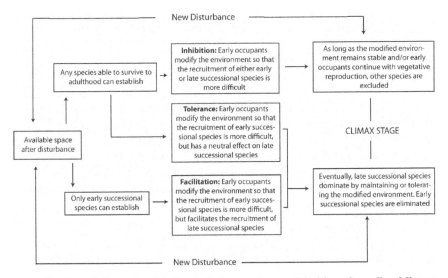

Fig. 3. The Connell-Slatyer model of ecological succession, modified from Connell and Slatyer (1977), and Vega Peña (2005).

composition after the disturbance and during the recovery process. Small gaps produced by the falling of large branches or single trees would favor the development of shade tolerant trees. Larger gaps produced by the simultaneous falling of several trees or other external factors such as fires, wind storms and landslides would favor light demanding or late tolerant species. Based on reproductive strategies, Denslow (1980) divided forest species into three general categories known as large-gap specialist, small-gap specialists and understory specialists.

At this point the question that arises is how important are all these strategies for forest management and silviculture? Considering that silvicultural treatments, including logging and extraction, often mimic gap openings aiming to promote desirable species (mostly by reducing competition), it is particularly important to know what and how gap size influence desirable species and how the structure of the forest is likely to change after the intervention. For instance, a logging system that generates small gaps would be advisable in areas where commercial species are shade tolerant or semi-tolerant whereas a logging system generating larger gaps would be tolerable in forests where commercially valuable long lived pioneers could dominate the gap afterwards.

Four important ideas should to be established at this point. If silvicultural intervention generates gaps, the latter should be considered substitutes of the natural gaps that randomly occur in the forest. It is likely that any intervention will increase the total gap area in a short period of time, so that the aggregated effect of the intervention in the medium and long term is hard to predict. Contrary to naturally occurring gaps, silvicultural interventions have a heavier impact on the forest soils when ground machinery is used and depending on the damage caused to the soil, the recovery process could be delayed or inhibited for a long period of time. The application of similar interventions over several growing cycles is likely to affect the so called climax stage of the forest and the continuous use of large gaps to promote regeneration affect negatively the competitive capacity of shade tolerant species and changes both the species composition and the ecological dynamics of the forest. Finally, the succession process is not predictable or unidirectional. There are several sources of unpredictability and most ecological processes are only partially understood. This makes some forests easy to model whereas others are just too complex to model at a reasonable cost (Taylor et al. 2009).

In spite of the latter, it is important to remember that the survival and development of any particular tree largely depends on the level of competition for limited resources, on the facilitation of competition provided by the architectural construction, on growth rates, and on the biotic interactions (Gómez-Pompa and Burley 1991). In addition, it should be noted that the classification of species into temperaments is an oversimplification of the complex changes experienced by species along their lifespan, either as part of their ecology or mediated by the interactions with the surroundings.

Silvicultural Systems

In a broader sense, silvicultural systems can be classified as monocyclic and polycyclic (Leslie 1977). In the first case, most if not all timber, is harvested in a large operation that creates large gaps. Regeneration is either natural or artificial (planted), and light demanding commercial species are favored. Several silvicultural activities can be applied in order to facilitate the growth of desirable individuals and the whole process may be repeated after the new tree population is commercially matured. In the case of polycyclic systems, desirable trees are selectively removed from the forest and medium to small size gaps are created. This strategy benefits shade tolerant commercial species and other silvicultural interventions may be implemented to reduce competition from desirable individuals. The next timber extraction could take place as soon as there are sufficient individuals above commercial size to financially justify the intervention.

Lowe (1977), Whitmore (1984), and Lamprecht (1989) offered descriptions of several silvicultural systems in both West Africa and South East Asia. Most of these systems are monocyclic in principle and depend on the strong presence of long lived pioneers of economic value (e.g., mahoganies in Africa and dipterocarps in Asia). However, there is much less experience with monocyclic systems in Neotropical forests. Whitmore (1991) suggested that many Neotropical forests are dominated by shade tolerant and slow growing species that do not benefit from larger gaps. For instance, in a 50 ha plot in Barro Colorado, Panama, Hubbell and Foster (1992) found that the abundance of species after a monocyclic intervention can be highly unequal and that mean density of particular species may be as low as two individuals per species per hectare.

In Latin America, there is growing experience with polycyclic systems (de Graaf 1986), but examples of monitored forests with more than two cutting cycles are rare. Whitmore (1984) and Lamprecht (1989) also mention examples of polycyclic systems in the Philippines and Indonesia. All these authors agree that there is sufficient technical experience and ecological knowledge for the sustainable management of tropical forests in all subcontinents. Unfortunately, the lack of economic resources to maintain national forestry services and inconsistencies in terms of policies, either by subsidizing competing industries or directly undervaluing forest resources, have hindered forest management in most areas. The Queensland management systems for natural and planted forests deserve a special mention at this point. After more than 100 years of forest management in the rainforests of north Queensland lessons have been learned from sporadic inadequate practices and today those experiences form the basis of a moderately sustainable use of forest resources. It should be mentioned that those lessons permeate both private and public forests. Vanclay (1996) offers a quick review of the system since 1873.

In Europe and North America, industrial forest management has promoted the homogenization of forest stands by implementing both polycyclic uneven-aged systems and monocyclic even-aged systems. This has been carried out by means of controlling fires and the use of both natural and artificial regeneration, which has generally favored coniferous over broadleaf trees. However, environmental concerns and the increasing discussion on ecosystem services during the last two decades are turning the balance in favor of uneven-aged systems which allow for greater diversity on managed forest stands (Graham and Jain 1998; Hagner 1999). Such practice resembles the German Plenterwald, which is used in several parts of Central Europe (von Gadow and Puumalainen 2000).

Alternatively, silvicultural systems can be further classified by partially following the typification of interventions suggested by Bawa and Krugman (1991). According to the latter, clearing, replacement, natural regeneration, and restoration systems are recognized. The following paragraphs offer a description and a selection of examples of silvicultural systems based on this classification.

Clearing Systems

In the presence of a significant number of mature trees of commercial or desirable species and under the assumption that natural or artificial regeneration will take place; or with the certainty that young trees and seedling from desirable species are present under the canopy, clearing systems can be recommended. Under these systems most, if not all of the valuable trees, are harvested whereas undesirable individuals or species are eliminated (Leslie 1977) by cutting, poisoning, or girdling.

If natural regeneration is assumed, good knowledge of the ecological principles governing the reproductive process of desirable species is indispensable. The harvesting of commercial trees and the elimination of undesirable individuals should be implemented in terms of timing and intensity in a way that best replicates the environmental factors that are necessary for the establishment and survival of desirable trees or species. Inevitably, some unwanted species or trees can establish themselves in the large openings created after harvesting, and weeding activities could extend for a few years until desirable individuals dominate the succession. Rotations are relatively large reaching up to 80–200 years in temperate zones, depending on growth rates (Carle et al. 2009) and the high demand for timber and inconsistencies in terms of policy and objectives have shortened rotation time in many managed areas.

Lamprecht (1989) refers to these as gradual transformation systems, since the final intended result is for the system to resemble a simplified ecological version of the original forest. One of the general goals of this type of forest management is the homogenization of the forest stand in order to reduce general costs and increase control on the growing stock. In classic forestry, most of the systems under this category fall into the concept of monocyclic management with more or less even aged stands.

Modern management of Indian forests initiated in the late 1800s with selective felling of Teak (*Tectona grandis*). However, due the large size of this country, several systems have been applied from the tropical regions to the Himalayas. Clearing or uniform systems were introduced in the 1920s to manage moist deciduous Teak and Sal (*Shorea robusta*) dominated forests. In both cases weeding and thinning activities were not fully implemented in part due to the high demand for wood during times of armed conflicts and the final result resembles a replacement system. Some forests were converted to agriculture, rotations were shorter over time and Teak plantations for natural regeneration were insufficient (Nair and Mammen 1985).

Another example of a clearing system is the Indian Irregular Shelterwood System (IISS) used to manage forests dominated by Sal, Deodar (*Cedrus deodara*), Fir (*Abies pindrow*), and other evergreen forests (Negi 1996). By dividing forests into sections or blocks with rotations of 120–200 years and establishing minimum commercial diameters above 70 cm, the system was created to maintain an uneven age forest mass that could keep most of the trees below 40 cm of diameter and the majority of mother trees after the initial felling (Negi 1996; Negi 2000; Appanah 1998). The first felling has the dual objective of harvesting commercial trees and eliminating competition for the future crop trees and further interventions, independent in each section or block, depend among other things on the regeneration needs of desirable trees (Negi 1996). Problems with natural regeneration have been documented as the main constraint of the system, particularly with *Shorea*, and planting seemed too costly (Appanah 1998).

Similarly, the Punjab Shelterwood System (PSS) was developed for the western Himalaya to manage Chir-pine (*Pinus roxburghii*). It is similar to the IISS with the particularity of creating six blocks where tree ages were arranged in periods of 20 years and are managed until harvestable trees reach 120 years of age. Interventions to eliminate competition were applied periodically according to the block age (Negi 2000). In Andaman Islands, the Canopy Lifting Shelterwood System (modified from IISS) was finally suspended in 2001 (UNESCO 2010) and substituted by a selective system. Under this modification, minimum harvestable diameter changed according to groups of species (hardwood, softwood, others), weeding is applied for the first year, climbers cut in the second and thinning took place at increasing intervals of 6-15-30-50 years (Nair and Mammen 1985).

Starting in the 1930s The Malayan Uniform System (MUS) was developed for Malaysian lowland dipterocarp forests in order to transform them into even age stands dominated by commercial species. The system included an initial elimination of climbers before main harvesting, the elimination of undesirable individuals six months later, and liberation treatments applied on the growing stock if necessary (Lamprecht 1989; Whitmore 1984; Matthews 1991). In this case, climbers were controlled as needed. Valuable trees below the minimum cutting diameter of 50 cm remained as growing mass for the next cutting cycle (Weidelt 1996). According to Schmidt (1987), MUS had acceptable results in the lowland forests, but was less successful in the upper dipterocarp forests. A modified version of MUS was then implemented in higher areas with one difference being the reduction in the poison girdling of remaining trees after felling (Whitmore 1984). Unfortunately, many of the forest areas where MUS was implemented were overexploited due to unrealistic economic expectations (Whitmore 1984) or cleared for agricultural expansion (Lamprecht 1989).

Apart from dipterocarp forests, the British developed in 1944 the Tropical Shelterwood System (TSS) for the Nigerian forests. The system was based on 20 years of experiences and the main goal was to achieve at least 100 desirable seedlings with a height of 1 meter per hectare before felling of commercial trees around year six (Schmidt 1987; Lamprecht 1989). The process started with the elimination of undesirable vegetation below a DBH of 5 cm along with the liberation of lianas and later continued with further liberations in an upward manner starting from the lower story. The idea was that if after the sixth year, regeneration did not reach the desirable goal, liberations and inventories would continue until the target was reached. After felling all commercial trees, the remaining forest would be cleared depending of the type of regenerating trees, light demanding or shade tolerant (Lamprecht 1989). After one million hectares were managed, the TSS was discontinued in part due to administrative incompetence and social pressures. However, modified versions of the TSS were later applied in Ghana and some selected forests in Nigeria (Neil 1981).

In the early 1950s, a top-down approach known as L' uniformisation par le haut (UPH) was experimentally introduced to the Belgian Congo.

After an initial inventory, the system basically consisted in the weeding and thinning of all undesirable trees and climbers in all canopy layers in addition to the felling of mature commercial trees (Neil 1981). Several thousands of hectares under management had a high percentage of commercial species that were also well distributed along all forest layers and initial results suggested that it was possible to homogenize several growing layers with valuable species (Lamprecht 1989). However, high implementation costs, low DBH increments, and the political uncertainty after the independence of the country affected the long-term stability of the experiments (Neil 1981; Schmidt 1987).

In Uganda and other parts of the French and Belgian ex-colonies, modifications of the TSS and UPH were tested over the years with rotations varying between 40 and 80 years. In Uganda, girdling and poisoning was reduced, whereas weeded trees were transformed into charcoal to cover some of the operational costs. In Sierra Leone and in poorly stocked forests under the French and Belgian administration, several forms of enrichment were tested in lines or patches with moderate success and high maintenance costs (Neil 1981). Most of the African experiences were undermined at the end of the colonial area and many post-colonial states did not have the technical capacity to manage or monitor the exploitation of their forest resources.

In Latin America, the best example of a clearing system is the so-called High Shade Shelterwood System (HSSS), which started as an adaptation of the TSS for easily erodible soils in Trinidad. Implementation started in the late 1930s in forests containing a large number of light-demanding species with commercial value (Neil 1981). After extracting all commercially valuable trees, larger undesirable ones were eliminated leaving an open canopy of small crowned trees for at least 18 months without any interventions in order to control erosion and facilitate regeneration. After this time, selection of desirable trees and tending of regeneration took place and valuable trees were divided as fast-growing or light-demanding and slower-growing or shade-tolerant. Around year five, remaining undesirable trees were eliminated and thinning took place after year ten (Neil 1981; Lamprecht 1989). The aim was to create a two layer forest with a rotation of 30 years for fast growing species and a rotation of 60 years for slow growing trees (Lamprecht 1989). Similarly as in Uganda, most of the residuals of commercial trees as well as undesirable trees were used for charcoal production and private charcoal makers were a very important part of the cost saving strategy. Around 2600 ha of forests were managed in this manner but the system was substituted with the Periodic Block System explained in the next section (Fairhead and Leach 2001; GoRTT 2011).

It is also important to mention that clearing systems have been successfully applied in many temperate and boreal areas with much lower growth rates. In Northern America for instance, clearing systems are normally applied on large industrial concessions. In Canada, the development of operating ground rules started in the late 1950s in the Alberta Forest Service with the testing of different modifications of clear cutting. Before that time, most concessions were based on cutting diameter limits for the sawmill industry and natural regeneration with different levels of success (Udell 2003). In recent times, different alternatives of clear cuttings based on strip or block approaches and shelterwood systems with or without assisted regeneration are available depending on the species composition (NRC 1995). In Eastern Canada, forests dominated by jack pine (*Pinus banksiana*), oaks, and cherries or black spruce (*Picea mariana*) in the boreal areas are normally managed under variations of the shelterwood system with particular emphasis on protecting regeneration (Groot et al. 2005; Cockwell 2012). The density of the forest stands can be reduced by 70% leaving several seed trees (Cockwell 2012). Clear cutting is still common in aspen-spruce mixed stands in the Western boreal forests, with different levels of retention of trees or patches of forest, and with a combination of natural and assisted regeneration depending on the conditions (Groot et al. 2005). The implementation of activities such as commercial and pre-commercial thinning, pruning, inventories, and Timber Stand Improvements (TSI) are performed at the discretion of the concessionary and provincial rules (NRC 1995).

In the Nordic Countries similar systems have had great success in a region where non industrial forest ownership is the norm. In these countries, policies, management, and industries have been aligned to compensate for long rotations. Policies facilitate the commercialization of timber and allow forest owners to have access to credits based on the expected value of their forest, whereas national infrastructure facilitates the movement of raw materials. Industries have been developed in order to utilize most of the biomass extracted during thinning and final felling and service companies have specialized in the management and tending of small forests stands. Revenues from forest and timber related taxation have been reinvested in institutions and programs in order to improve the sector. In spite of that, concerns remain over the state of managed forest areas when compared to natural forests but the contribution of the former on the economic development of these countries has been significant. If product provision as well as significant ecosystem services are also considered in the discussion, it can be noted that managed forest have supported more than the commercial aspect of the timber industry.

Generally speaking, clearing systems have lost relevance as an ideal model for tropical regions in modern times and this type of management has been gradually degraded or has disappeared altogether due to unrealistic economic and productive expectations or as a product of sustained socioeconomic and political pressures. The three main negative effects on the systems have been the long waiting periods between rotations, the heavy impact on natural ecosystems, and social interactions due to the magnitude of the interventions (Weidelt 1996). Seydack (2000) offered additional reasons for the decline in the implementation of clearing systems. Clearly, there is a size constraint when implementing clearing systems. In order to sustain economic viability it is important to maintain large enough areas and at different stages of management in order to generate a positive cash flow from occasional logging and commercial thinning. At the same time, the extent of the managed are as usually under the frame of private concessions tends to affect communities that depend on forest resources or live within or close to the concession forest.

Replacement Systems (Conversion Systems)

In contrast to the slower transformation of the structure and composition of the forest stands in the previous category, a replacement system implies the elimination of the forest stand with the exploitation of commercial species and the replacement within few years with either introduced or native commercial species (Gómez-Pompa and Burley 1991). Depending on the region, commonly introduced species are Pines, Eucalyptus, Teak, and Gmelina (*Gmelina arborea*). Large scale plantation forestry is not described in detail in this chapter and the emphasis is still placed on systems using a mixture of species and on the natural interactions presented within natural forests.

These replacement systems also include several forms of shifting cultivation practices if forest cover is to be re-established (Gómez-Pompa and Burley 1991), tree integration with intensive agriculture is desired (e.g., cacao, coffee, pepper, some hot peppers), or agroforestry systems implemented (Lamprecht 1989; Gómez-Pompa and Burley 1991). An example of the latter is the Taungya, a system where tree cover remains after the agriculture production is discontinued. Examples of replacement systems are rare in the most developed parts of the world as conversion of forests into other uses is heavily regulated and even agroforestry is practiced in areas that have been without forest cover for a long time.

Shifting cultivation is probably the oldest form of agriculture and it is based on an alternated and periodic use of land. Plots under crops are cultivated for a limited period of time and followed by a lengthy period of soil recovery. Later production is shifted to a new plot and a successional mosaic of small patches is formed at the landscape level over time. Historically, shifting cultivation was practiced all over the world, including boreal areas and several authors have treated shifting cultivation as an evolutionary step into agriculture which relied on more intensive practices (Pelzer 1948; Popenoe 1959; Spencer 1966; Ruthenberg 1971). At present, this productive concept is mostly practiced in marginal areas of the humid tropics or low-populated savannas where millions of small farmers are still implementing variations of shifting agriculture as it still remains a profitable system and often the only one available (Dove 1983; Nielsen et al. 2006; Lininger 2011).

It was suggested long ago that population pressure and the subsequent shortening of the recovery period, would render shifting cultivation unsustainable and that technical improvements were needed in order to sustain and increase productivity. The utilization or introduction of woody perennial crops had been seen as one of the alternatives (Lafont 1959). Under the bush-fallow system, traditionally used in Melanesia, farmers continue the management of plots under fallow to obtain other food items and eventually woody products before they come back and re-establish new crops (Barrau 1959). Modifications of this idea have been introduced in many areas under the general frame of improved fallows where multipurpose trees are introduced to diversify production and accelerate the recovery of soil productivity (e.g., with nitrogen fixing species). The resulting landscape is a mixture of young fallows and areas of secondary forests at different successional stages, with little or no examples of late forest successions.

When perennial crops with better market value, such as rubber, coffee or cacao, are introduced, shifting plots can remain productive for longer periods and this can be considered as a longer form of shifting cultivation. In Jambi, Indonesia, rubber was introduced in 1904 and farmers planted it in shifting cultivation fields mixed with other crops. This system is locally known as hutankaret. Due to the profitability of rubber, many farmers maintained several of their trees and managed their plots in the form of jungle rubber with the rejuvenation of the plot occurring when rubber productivity declined after 25 years or more (Joshi et al. 2002). In the lowlands of Papua New Guinea, a similar system based on coffee has developed since the 1950s. During the cultivation period, fruit trees were introduced and after crops were eliminated, coffee and *Leucaena* were incorporated to the fallow, remaining productive for up to 30 years (Allen 1985). In these systems, it is common not to eliminate young individuals of valuable timber species during the initial clearing or during regeneration for the maintenance of shading and due to the additional economic value of those species at the end of the cycle. In this mixed productive landscape, some plots are old and diverse enough to be compared to secondary or advanced successional forests before they are cleared again. Under the right circumstances, these plots can become permanent managed units with a multilayer forest cover known as homegardens and can be seen as complex agroforestry systems when trees are incorporated and managed.

Some of the most cited and studied examples of this type of homegardens are from Indonesia (Wiersum 1982), but similar examples can be found in many other countries in Southeast Asia, Africa, and Latin America, particularly in areas with high population density (Fernandes and Nair 1986) where shifting cultivation is unsustainable. The Kebunkaret or rubber-gardens in Kalimantan are more intensively managed land units than the jungle rubber, with a more systematic use of other tree species for timber, firewood, and other products as well as the use of diverse shade-tolerant crops (Ihalainen 2007). In Southeast Mexico, homegardens are subdivided in well differentiated units with the yard around the household being mostly cleared of trees and used for complementary agriculture activities, the garden being densely planted with ornamental, medicinal, and seasonal plants and the orchard, covering most of the area with a combination of natural regenerated and planted trees and shrubs for very specific uses (Alvarez-Buylla et al. 1989). In most of these cases, homegardens have a tendency to simulate the structure of a forest with several canopy layers, high biodiversity, and complex species interactions (Fernandes and Nair 1986).

Apart from the mentioned replacement systems, the incorporation of trees into agricultural oriented productive schemes, there are also examples of agriculture brought into forestry systems, particularly plantations. For instance, the systematic implementation of the Taungya System started in 1868 in Burma for the production of teak and rapidly expanded along South East Asia, Africa, and Latin America (Wiersum 1982). Unreliable statistics indicate that up to two million hectares in Southern Asia and one million in Africa have been planted under this system (Lamprecht 1989). Initially, *Taungya* had the dual objective of reducing shifting cultivation by offering farmers temporary agricultural rows between tree ones. This system is known as alley cropping and was strictly directed in favor of the tree component due to the lower value of plant species in overpopulated areas with land shortages (Grandstaff 1978; Wiersum 1982).

Several variations of the system were developed later on, but the basic principle included the elimination of the original vegetation to later incorporate some form of agricultural production during the early years of the plantation until the crown cover closed. Over the years, the main objectives of the system changed and became less exploitative, with better arrangements for local communities, and with a more diverse selection of tree species including shorter rotations and multiple uses (Grandstaff 1978). Since the 1970s, the Forest Service in Indonesia introduced an intensified version of Taungya for landless farmers in selected areas with poor soils and moderate population pressure. Under this system intercropping is allowed for a few years in a mixed plantation of Teak and *Leuaena leucocephala*, a nitrogen fixer used for mulching 2-3 times per year. The use of improved crops and fertilizers is supported by the Forest Services. Just in the island of Java, 40,000 ha of teak were established this way between the late 1970s and early 1980s (Wiersum 1982). *Gmelina arborea, Shorea robusta, Terminalia superba, Cupressus* spp., and *Pinus* spp. are examples of other species established using Taungya in Asia and Africa (Lamprecht 1989). As with many other systems before, Taungya has a mixed history of failures and successes in most cases not related to silviculture. The right combination of benefits and responsibilities for farmers, an adequate selection of species, and monitoring capacity, are key elements in the establishment and management of forested areas using this approach.

Natural Regeneration Systems

Also referred to as selective cutting or selective system. In these systems only few commercial trees are extracted during each cutting cycle for timber volume in the forest to remain more or less stable in the long term. However, moderately decreased commercial volumes have been observed in several forest areas after the first intervention. In this system it is also assumed that the felling pattern simulates naturally occurring gaps to some extent and that regeneration of desirable species develops within the gaps created by the felled trees.

Gómez-Pompa and Burley (1991) identified several groups of species that could benefit from the creation of gaps and herein these groups have been reduced from five to four. It is important to remember that the mechanisms by which these groups benefit depend on the size of the gap, soil conditions, and the intensity of the damage.

1. Species that are already established in the site and that managed to survive the mechanical damage and environmental changes (Martínes-Ramos 1985 cited by Gómez-Pompa and Burley 1991). In this group are included the light demanding species, the survivors from previous gaps, and those dormant species ready to react to the new microenvironment.
2. Species that are present in the seedbank of the topsoil. These species come out stimulated by changes in light, temperature, and mechanical factors (Whitmore 1983 cited by Gómez-Pompa and Burley 1991).

3. Species growing out of propagules surrounding the gap (Peñalosa 1985 cited by Gómez-Pompa and Burley 1991).
4. Species migrating into the open area during the time the gap exists (Gómez-Pompa and Burley 1991). These species are not present in the affected area before the creation of the gap, but rather arrive as a consequence of it.

As many of the species colonizing and later developing in the gaps are neither commercial nor desirable, situations arise when assisted regeneration, identified as the reduction or elimination of competition affecting desirable species or individuals and directed regeneration, defined as the introduction of desirable species or individuals are performed. Assisted regeneration often includes the weeding, killing, or thinning of competitors. The killing without removal reduces competition without creating additional gaps and minimizes the risk of mechanical damage to desirable individuals. Directed regeneration is normally carried out by introducing individuals on systematic patterns that are usually lines or aggregates.

Unfortunately in tropical and subtropical areas, there are only limited examples of natural regeneration systems with long term sustainable production of timber. Some experimental areas have been abandoned due to lack of resources or political tensions, whereas many other productive areas have become a one-time extraction operation where little has been done to ensure the economic and environmental viability of future extractions. In the worst cases, forests have been re-harvested before recovering under the excuse of silvicultural activities, thus further eroding their economic and ecological value.

In Malaysia, building on the experience of the Malayan Uniform System and working on already logged areas, the Sarawak System was initially implemented in Borneo. The system included liberation thinning that is not carried out systematically but rather on desirable trees affected by direct competition. Selective logging accounted for just a fraction of total commercial volume (Schmidt 1987). In order to prepare the thinning, an initial inventory was applied to identify the leading desirable trees and their level of direct competition (Matthews 1991). The Sarawak System is one of the most successful systems still applied in Malaysia today (Schmidt 1987; FDS/ITTO 2012).

Similarly, the Philippine Selective Logging System (PSLS) was practiced since the late 1950s (Whitmore 1984), with felling quotas of 30% for trees between 15 and 65 cm, 60% for trees between 65 and 75 cm, and rotations periods between 30 and 45 years. The methodology included a pre-felling inventory of commercial trees above 15 cm of diameter as well as an inventory of residuals after logging operations. Marking of trees to be harvested and felling direction, marking of desirable residual trees, and further stand improvements after 10 years (Lamprecht 1989) were also carried out. Management in old growth forest was discontinued in 1992 after forest cover dropped below 20% and management has since been limited to residual forests areas (ITTO-FMB/DENR 1995). As a response, site specific variations of the Timber Stand Improvement (TSI) have been applied on young residual forests. Under this system, undesirable trees in direct competition with Potential Crop Trees or PCTs are eliminated by girdling or cutting, thus enhancing the development of valuable trees and the quality of the forest in general. Experimental areas show positive results, but other areas show a mixed picture. In most cases unsatisfactory results are more related to administrative deficiencies than technical limitations (ITTO-FMB/DENR 1995).

In less diverse forests then the Filipino ones, the Indonesia Selective Logging System (ISLS) was applied with a fixed rotation of 35 years that concentrated on fewer species and the extraction of up to 15 trees per hectare (Lamprecht 1989). The system was simple and easy to monitor, but still had a short rotation considering the damage caused by mechanical harvest on the residual stock. The ISLS was substituted in 1989 by The Indonesian Selective Cutting and Planting System (TPTI) with later modifications in 1993 (Redhahari et al. 2002). Despite the introduction of pre-harvest inventories, residual inventory, enrichment, liberation treatments, and thinning, recent research suggests that the fixed 35 year rotation based on weak assumptions of productivity and the damage to the remaining mass is still unsustainable and affecting long term timber production. Extraction remains also high with at least 10 trees per hectare (Klassen 2002; Sist et al. 2003; Iskandar et al. 2006; Priyadi and Kanninen 2009).

In Africa, the improvement of stands or amélioration des peuplements (APN) was a system implemented in Ivory Coast that lied between a natural regeneration and a clearing system depending on the intensity of liberation. Experimentation started in the 1950s initially in logged stands and later in primary forests (Dupuy et al. 1999). After the inventory of desirable species, lianas were cut and a few months later, large undesirable trees were taken out of the forest. Undesirable trees with large crowns were poisoned in all canopy layers. The result was an almost random allocation of large and small gaps that controlled the establishment of light demanding species while favoring the growth of the remaining marketable species. Periodical interventions in the next 10 years maintained low levels of competition for the desirable trees (Lamprecht 1989). The system was discontinued in the early 1960s due to poor results; however, variations of the system have been included as part of the Sectorial Forestry Plan during the period 1988 to 2015 in Ivory Coast as well as in other projects in Cameroon (Dupuy et al. 1999).

In the Americas, experiences with natural regeneration systems are much more recent. In some countries and for several decades the extraction of selective species was the norm (e.g., *Swietenia* spp., *Carapa* spp., *Mora* spp., *Cedrela* spp.), with little attention to management or sustainability. Many timber industries have obtained most of the raw material at a very low cost, after forests were cleared for agricultural expansion and industries have moved with the agricultural frontier.

In Surinam, silvicultural experimentation started as early as 1904, but many sites were abandoned before 1925. During the 1950–60s and following the Dutch experiences with monocyclic systems in South East Asia, experimental areas were logged, lianas were controlled, and undesirable trees poisoned. Preliminary results suggested annual diametric increases of 1 cm for desirable species, but at a heavy economic cost (FAO 1993). After several small scale experiments in the late 1960s by The Centre for Agricultural and Forestry Research (CELOS), three new experimental areas were established to evaluate the previous results at a practical scale between 1975 and 1978. The optimization logging intensity and silvicultural treatments in those experiments is what is known today as the CELOS Management System or CMS (Hendrison and de Graaf 2011). CMS was divided into two components, the CELOS Harvesting System or CHS which implemented reduced impact logging and the CELOS Silvicultural System or CSS which implemented two to three targeted liberations of potential crop trees in order to keep costs low. After a forest inventory preliminary assessment, a management plan was prepared detailing the road network and the division of operational areas. A full stock-survey was then implemented where terrain conditions were also recorded and commercial logging oscillated between 20–30 m^3 per hectare (4–7 trees). The end result has been a detailed operational map (1:5000) showing forest variability, natural elements, target trees, trails and roads, as well as other relevant infrastructure or features (Zagt 2004). In Surinam, CMS has not been implemented outside experimental areas, but it has been included as part of the alternatives to manage national forest areas (Hendrison and de Graaf 2011). However, variations of the CELOS system have been implemented in Brazil, Costa Rica, Bolivia, Cameroon, and Ghana.

In Brazil, after a series of experimental treatments carried out in the 1950–60s, the government in cooperation with FAO initiated a multiple use forest project in the Tapajós National Forest in the Amazon basin (Schmidt 1987). By 1983 the trials with extractions of up to 16 trees per hectare proved to be too aggressive for a 30 year rotation. Starting in 1982, new experiments were established to evaluate the complementary elimination of 30–70% of the basal area remaining after all commercial trees were logged. A second alternative was a smaller extraction of only six trees per hectare and the elimination of 1/3 of the remaining basal area (FAO 1993). Based on these experiences and others in the region, the current Brazilian forest management is based on the Sustainable Forest Management (SFM) principle. As such, many of the CMS recommendations with a maximum harvest quota of 30 m^3 per hectare, full inventory of commercial trees, targeted liberations, RIL, and cutting cycles between 25–35 years depending on the forest type are followed in Brazil. In addition, a requirement of protection for 10% of all commercial trees per species and a mechanism for the chain of custody of harvested trees is also enforced in that country (Blaser et al. 2011).

After a couple of centuries of working and experimenting with clearing systems that facilitated a large conversion of mixed forest into even-aged conifer-dominated stands, Europe formally reintroduced the idea of selective systems as part of the toolbox for forest management during the late nineteenth century. Selective cutting was still common in small privately owned forests or communal forests along the continent (Troup 1928; Spiecker et al. 2004). This was carried out in response to observed reductions of growth in some forest stands, managed under clear cutting and shelterwood systems (Troup 1928). The idea gained further traction after the discussion of alternative silvicultural methods by Henri Biolley in 1901 (Schlaepfer and Elliott 2000; Schütz et al. 2012) and the description of the permanent forest or Dauerwald by Alfred Möller in 1920 (Troup 1928). Those two ideas fuelled the renaissance of the selective forestry or Plenterwald in Germany, France, and Switzerland as a strategy towards a more natural and sustainable forest management.

In modern times in Europe, the concept of Plenterwald has been expanded into the broader idea of Continuous Cover Forestry (CCF) to account for all potential goods and services provided by managed forest stands. CCF is not a sylvicultural method, but rather a management idea where natural processes and structures are maintained or even restored under a more or less permanent forest cover that allows selective extraction of trees. Even in the Nordic Countries, CCF is gaining public interest under the more strict nomination of uneven aged forest management.

In Canada, selection systems were introduced in mixed forest stands following the European experiences (Udell 2003), but with a more industrial approach. The selection was based on single trees, groups of trees in patches or strips (similar to a shelterwood system) and the aim was to maintain an uneven aged forest stand with strong dependence on natural regeneration. These methods have been applied in mixed boreal forests as well as in both Eastern and Western coastal forests, especially on old growth forests open for logging with different retention levels for old trees (Groot et al. 2005).

Restoration Systems

Despite being in the shadows for a long time, restoration systems have gained relevance in recent years (ITTO 2002; Lamb and Gilmour 2003; ITTO/IUCN 2005; Lamb et al. 2005; Rietbergen-McCracken et al. 2007) and have been seen as an alternative to complement the provision of goods and services coming from more natural forested areas that cannot be easily supplied by forest plantations. There are not well defined silvicultural activities since for these systems, the common starting point is a degraded environment (Gómez-Pompa and Burley 1991). Millions of hectares of heavily degraded natural forests, even aged industrially managed forests, poor secondary forests, abandoned agriculture fields, *Imperata* grasslands, and man-made savannas are potential areas to be managed as restoration systems.

ITTO (2002) suggested that by the year 2000 there could be over 500 million hectares of degraded primary and secondary forests in addition to the 350 million hectares deforested since 1950. Around 500 million people depend on these degraded or modified forests for their livelihoods (ITTO/IUCN 2005). A significant fraction of these degraded forests are part of a mosaic of land uses, that often include patches of more natural residual forest as well as patches at different stages of degradation (Lamb et al. 2005). Many of these areas have the potential to be managed under the forest restoration framework and the capacity to recover, at least partially, their potential to supply goods as well as ecosystem services. In countries like Costa Rica, Philippines, Sri Lanka, Nepal, and some Indonesian islands, degraded and secondary forests represent the main sources of forests goods and services (ITTO 2002).

The level of degradation can be moderate as in a forest where the most valuable species have been extracted but where forest structure and functions remain severe when significant tree cover has been lost, top soil is missing or disappearing, and the area has reached a point where natural regeneration processes are inhibited. In some cases, grasses or other invasive species can dominate the degraded area and create a new ecological equilibrium preventing the return of forest cover. ITTO (2002) indicated that degraded forests are those in which the natural structures, processes, functions, and dynamics are altered to the point that makes medium term recovery difficult and the short-term resilience capacity of the system has been compromised.

As many of these areas have partially lost their economic potential and therefore become less attractive for industrial activities, restoration tasks often include collaborative management with local stakeholders to deal with the dynamic and complex interactions between people, natural resources, and land uses (ITTO/IUCN 2005). In most cases, the priority is to stop the drivers of degradation in order to restart the natural processes necessary for soil and vegetation restoration, with fire protection and erosion control as common primary goals. In some cases, assisted generation is needed by introducing desirable species in gaps or by establishing a tree cover to nurse natural regeneration. Nursing species can be either native trees like some cases in Costa Rica (Haggar et al. 1997; Cusack and Montagnini 2004) or exotic species like some examples from Ethiopia (Yirdaw and Luukanen 2003; Alem and Pavlis 2012).

With low to moderate degradation, management of primary or mature forests includes the control of the drivers of degradation as well as silvicultural activities targeting the recovery of species or natural processes. In that way, the control of unsustainable practices by introducing improved planning and techniques and the control of unnatural fires are implemented (ITTO/IUCN 2005). Such activities could also include liberation-thinning, reintroduction of species, reduction of the populations of species favored by the degradation processes, management of natural regeneration, and other case-specific activities. Many of the actions available are similar to the ones implemented in natural regeneration systems to increase the commercial value of the stands. Under restoration systems, additional value can be given to ecologically important species or species with economic value other than timber.

In areas where forest degradation is severe but where soil degradation still allows for the natural regeneration of trees, forests can be re-established if the drivers of degradation are controlled or eliminated. Examples of these areas are abandoned agricultural fields, low productivity grazing pastures (ITTO/IUCN 2005), heavily depleted forests, or areas recently cleared by human or natural forces. For the natural regeneration to take place, vegetative material or seeds have to be available either from the seed bank or from nearby patches of forest. However, the quality and diversity of the regeneration sources does have an effect on the regeneration process. As mentioned before, fast growing forest plantations can be used as nursing cover for other species and a strategy intended to accelerate soil formation, and simulate early successional conditions (Lamb et al. 2005).

Under severe soil degradation, besides controlling drivers of degradation, priority should be given to restoring and maintaining primary processes such as nutrient cycling and hydrological regimes that eventually facilitate the appearance of microhabitats for the natural germination of seeds (ITTO/IUCN 2005). A common mistake has been the introduction of trees in soils that are poorly structured to sustain them. As before, nursing species can be used to accelerate the restoration of primary processes and these could include small herbaceous plants or trees that tolerate very poor soil conditions similar to the previous case. If the land has been overtaken by grasses then biological, mechanical, or chemical control is likely required in order to break the ecological cycle of the grasses. The introduction of cover crops or trees, fire control, the mechanical elimination, and deep tillage are the most common alternatives to reduce the competitive advantage of grasses. *Imperata cylindrical* is probably the best known example of an invasive grass that dominates areas after human disturbances in large parts of Southeast Asia, Africa, and particular habitats in the Southeastern United States (MacDonald 2004).

When a new forest grows from cleared or almost cleared areas, the resulting ecological unit is known as a secondary forest. These units can be managed for ecological restoration when the objective is to replicate as much as possible the original forest cover or they can be managed to favor the development of a particular set of goods or services with economic or recreational values. However, experiences from Costa Rica suggest that early interventions in secondary forests are ineffective due to the strong dynamism of early succession and silvicultural activities after 15 to 20 years of development can be better targeted as the forest develops a more stable vertical and horizontal structure (Monge 2000). The management of secondary forests follows most of the same principles as the management of primary forests and over time many secondary forests could resemble the structure and floristic composition of a primary forest.

In order to facilitate the restoration process, it is important to realistically balance the trade-offs and the elements such as biodiversity conservation, economic development, and cultural values as win-win situations are rare (Lamb et al. 2005). In most cases, the success of restoration efforts is more related to social, political, and legal issues rather than to technical limitations (ITTO/IUCN 2005). Another common difficulty is the misdiagnoses of the drivers of degradation (ITTO/IUCN 2005). Erosion, seedling predation, fire, or overgrazing are easily identifiable, but other elements such as seed predation or the disappearance of microhabitats for the regeneration or survival of key species are more difficult to recognize. Savory (1969), suggested that overgrazing by large game (mostly elephants) followed by fires, were the main driver behind the ecological collapse of certain areas in Zimbabwe. Game management was introduced to control populations, but degradation continued. Later, Dublin et al. (1990) indicated that the interactions between fire and elephants were far more complex and that the main driver of degradation changed between locations, depending on local conditions. In spite of the latter more intricate considerations, hundreds of elephants were eliminated as part of the restoration strategy.

It is worth mentioning that agroforestry is easily recognized as one of the restoration systems when implemented in degraded areas or in marginal agricultural lands in order to sustain or recover the productive capacity of the soils by introducing trees to farms. The objective is not to restore forest areas, but rather to complement or facilitate other restoration activities by decreasing the need of opening new agricultural plots, reducing pressure on forests by providing goods and services closer to the farms, sheltering seed dispersing animals, and providing a softer land use around recovering forests when compared to traditional agriculture or pastures. Contrary to some of the replacement systems mentioned before,[1] the starting point here is a low productivity agricultural field where trees are incorporated into the system to control degradation processes or to include complementary elements into the system such as an increase in shade and enhanced recycling of nutrients.

In several European countries, the transformation of even aged conifer forests into uneven aged mixed broadleaves forest stands is considered a form of rehabilitation management based on low-impact and frequent thinning operations (Stanturf and Madsen 2002). Among Nordic Countries, prescribed fires, allowances of coarse woody debris, restoring aquatic systems, and management of key non-tree species like moose and beavers have been introduced in order to restore intensively managed forest stands into more natural units (Similä and Junninen 2012). Afforestation, defined as recovering tree cover on abandoned agricultural lands, is also an important restoration activity in some Eastern European countries after marginal agricultural lands became unprofitable. Their implementation has a strong government support and includes both natural and assisted regeneration of mostly broadleaves species (Stanturf and Madsen 2002). Similar examples can be found in other advanced economies where the intrinsic value of forests is affordable for societies.

Much has been documented in terms of positive and negative experiences in the last 20 years. The references presented herein in relation to restoration systems, are just a small part of the literature available worldwide. It is worth pointing out that with the right political framework a significant proportion of the current degraded areas could be put under forest restoration regimes. This is particularly important considering the role that degraded and secondary forests play on the provision of ecosystem services and the maintenance of livelihoods for millions worldwide.

Closing Remarks

The evolution of silviculture has been a long learning process, not only technically, but also at a more philosophical level. It started with the

---

[1] Some variations of the *taungya* system classify as restoration management if the starting point is a degraded land and the underlying goal is to restore natural forest cover.

realization that forests were a decreasing resource after the selective and unsustainable exploitation of valuable trees. After that, the control over the utilization and maintenance of forested areas due to pressure and demand on the resources increased the maximization of profits and volumes. Then the idea of domestication and homogenization of forest stands for economic optimization did not meet the expectations of forest managers and societies demanded forest utilization to focus on non-timber products. Finally, in recent years we are coming to the realization that the "ideal" management of forest stands requires a more or less permanent forest cover and a certain level of naturally occurring processes as a proxy for resilience. The complete process has not been in any way an isolated task but rather the reflection of changing socioeconomic views and philosophies, including of course how we understand our relationship with nature. My personal opinion, just like the view of many other forest managers, is that the technical knowledge to manage forest stands in a continuous manner has been available for quite some time already and that despite our incomplete ecological knowledge at certain levels, the ultimate aim of sustainable forest management is very much feasible at this point in history.

However, as with many other issues affecting the forestry sector, forest management is mostly limited by administrative and economic shortcomings rather than by the lack of detailed ecological knowledge. Conflicted legislation, complicated and costly bureaucratic requirements, and unreasonable expectations of short term financial returns are undermining the long term productive potential of many forest areas. In terms of legislation, the first step is to eliminate perverse incentives that promote deforestation or forest degradation by undermining the real value of forests or subsidizing alternative uses. Additionally, it is important to create the right legal framework to secure access or ownership over resources, to facilitate technical assistance in the preparation and implementation of management plans, and to promote those markets for goods and services derived from new and existing forested areas.

Countries like Finland and Sweden are examples of how the right combination of policies, including the industrial sector, can sustain a large and productive forest cover. Even if well managed forests do not provide the same quality of ecosystem services as less intervened areas do, this is probably the best alternative if society cannot afford the opportunity cost of a comprehensive network of protected areas. Costa Rica is another example of how the right combinations of policies, early subsidies, and market incentives have allowed the total forest cover to increase. Globally, there are plenty of reasons for optimism as recent UN conventions brought attention and economic resources to deal with the problem of unsustainable forest management.

The increasing adoption of Criteria and Indicators (C&I) for sustainable forest management and certification since their introduction in 1992 at the UNCED in Rio, tailored to local or regional realities, is a very promising trend. More countries are adopting and incorporating globally recognized C&I (e.g., FAO, ITTO, CIFOR, FSC, PEFC) into their own national legislation. For instance, Canada, Brazil, Finland, Australia and Costa Rica have already incorporated C&I as part of their forestry legislation while similar initiatives are ongoing in several countries as part of nine regional initiatives promoting C&I globally (Günter et al. 2012). It is also important to mention that the global C&I initiative strongly overlaps with the ecosystem approach for managing natural resources promoted under the UN Convention of Biological Diversity, as both aim to balance environmental, social, and economic concerns as well as the rights of future generations (Wilkie et al. 2003).

The Forest Law Enforcement, Governance and Trade (FLEGT) program by the European Union, supports improvements in the forestry sector of partner countries through its Voluntary Partnership Agreements (VPAs) as a way to reduce illegal logging using timber trade as an incentive. Other global institutions like the World Bank, ITTO, and FAO have been supporting the program and much has been done in terms of improving legislation and governance in developing countries and Eastern European nations. Progress is slow, but having adequate policies, guidelines, and regulations are the first steps towards sustainability in the forestry sector.

As mentioned earlier in this chapter, the inclusion of forest management as part of the mitigation activities under the UNFCCC continues to have a significant impact in the way we do forestry. As part of the Reducing Emissions from Deforestation and Forest Degradation (REDD+) initiative, much has been done in terms of improving legal frameworks, introducing new monitoring and management technologies, involvement of stakeholders, and in general, in terms of opening the discussion about what we want from our forests and on the different factors undermining sustainability in the sector. None of the global trends mentioned here income without criticisms and corrections and improvements are incorporated on a regular basis. In this way, it will still take time, but change is happening and we are moving in the right direction.

On the technical side, most of the management systems mentioned before rely on optimistic assumptions about natural regeneration and the survival of desirable species. However, more information is needed about the ecological requirements for the establishment and subsequent life cycles of several key tree species, especially for those of particular economic and ecological significance. In addition, more information on biomass productivity for target species under different levels of competition and management regimes will facilitate the estimation of more sustainable harvest quotas and cutting cycles.

In the tropics, it is reasonable to assume that quotas of up to 30 m^3 per hectare (seven trees or less) with rotations no shorter than 25 years can sustain productivity on biologically diverse forest stands. In more homogenous stands, the combination of dominant species may dictate the intensity and frequency of interventions with even aged stands as an alternative. Nonetheless, sustainability cannot be achieved without the compulsory implementation of Reduced Impact Logging (RIL) techniques, and in some cases Timber Stand Improvements (TSI), not to mention regularly monitoring of the growing stock as it is carried out in many advanced economies with strong forestry sectors.

Considering the ever growing expectations on the provisions of ecosystem services from forest areas such as carbon sequestration and food security, the continuous update of the integrated ecological knowledge on trees is vital for the management of forest stands. How forest management affects the provision of ecosystem services, should ultimately feedback into the way forest management is implemented. It is worth mentioning that the contribution of forest management to forest conservation is related to the maintenance or enhancement, up to a certain degree, of natural processes and in some cases to the maintenance of taxonomical and structural diversity. However, forest management is an intrusive activity and expecting the same ecological stability as in protected areas is unreasonable.

As indicated by ITTO (2002), forest management needs to be applied and understood within the framework of constantly changing conditions, at the landscape level, due to the changing needs of local and international societies. The involvement of most if not all stakeholders as part of the decision-making process cannot be underestimated. Many forms of small scale forest management, done by small farmers or communities, are usually overlooked as part of the larger national economies and often do not receive the necessary political or technical support needed to ensure long term sustainability. Nonetheless, small scale forest management can have an important impact on the local economies where it is implemented.

Acknowledgements

I would like to express my gratitude to the Viikki Tropical Resources Institute (VITRI) and the University of Helsinki for allowing me to allocate time and resources to work on this project. In a similar way, my gratitude to the Finnish Cultural Foundation for their support during the preparation of the document. I would also like to thank Prof. Markku Kanninen for his support and all the discussions about the state and future of forest management, many of his views have influenced parts of this chapter; my gratitude also to Emer. Prof. Olavi Luukkanen for sharing some of his classic forestry literature with me for the preparation of this chapter. Finally I would like to thank my old friend and Professor Ruperto Quesada for introducing me to the problematic and interesting world of tropical silviculture.

References

Agnoletti, M., J. Dargavel and E. Johann. 2009. History of forestry. pp. 1–28. *In*: V. Squire (ed.). The Role of Food, Agriculture, Forestry and Fisheries in Human Nutrition, Vol. II. Eolss Publishers, Oxford.

Alem, S. and J. Pavlis. 2012. Native woody plants diversity and density under *Eucalyptus camaldulensis* plantation, in Gibie Valley, South Western Ethiopia. Open Forestry Journal 2: 232–239.

Allen, B.J. 1985. Dynamics of fallow successions and introduction of robusta coffee in shifting cultivation areas in the lowlands of Papua New Guinea. Agroforest. Syst. 3: 227–238.

Alvarez-Buylla, M.A., E. Lazos and J.R. Garcia-Barrios. 1989. Homegardens of a humid tropical region in southeast Mexico: an example of agroforestry cropping system in a recently established community. Agroforest. Syst. 8: 133–156.

Appanah, S. 1998. Management of natural forests. pp. 133–149. *In*: S. Appanah and J.M. Turnbull (eds.). A Review of Dipterocarps, Taxonomy, Ecology and Silviculture. CIFOR, Bogor, Indonesia.

Barrau, J. 1959. The "bush fallowing" system of cultivation in the continental islands of Melanesia. pp. 53–55. *In*: C. Ratanarat, C. Tongyai, W. Brink, N. Tongyai, D.V. Sassoon, C.F. Siffin and I. Farrar (eds.). Proceedings of the 9th Pacific Science Congress of Pacific Science Association, Bangkok, 1957, Vol. 7.

Baur, G.N. 1964. Rainforest treatment. *In*: Unasylva (72), An International Review of Forestry and Forest Products. FAO, United Nations. Vol. 18(1).

Bawa, K.S. and S.L. Krugman. 1991. Reproductive biology and genetics of tropical trees in relation to conservation and management. pp. 119–136. *In*: A. Gómez-Pompa, T.C. Whitmore and M. Hadley (eds.). Rain Forest Regeneration and Management. Man and the Biosphere Series, UNESCO and the Parthenon Publishing Group, Paris.

Blaser, J., A. Sarre, D. Poore and S. Johnson. 2011. Status of Tropical Forest Management 2011. ITTO Technical Series No. 38. International Tropical Timber Organization, Yokohama, Japan.

Budowski, G. 1970. The distinction between old secondary and climax species in tropical Central America old forests. Trop. Ecol. 11: 44–48.

Budowski, G. 1984. The role of tropical forestry in conservation and rural development. The Environmentalist 4: 68–76.

Carle, J.B., J.B. Ball and A. Del Lungo. 2009. The global thematic study of planted forests. pp. 33–46. *In*: J. Evans (ed.). Planted Forests: Uses, Impacts and Sustainability. CAB International and FAO, Wallingford and Rome.

Cockwell, M. 2012. The forests of Canada: a study of the Canadian forestry sector and its position in the global timber trade (working draft). Limberlost Forest & Wildlife Reserve. Toronto.

Connell, J.H. and R.O. Slatyer. 1997. Mechanisms of succession in natural communities and their role in community stability and organization. American Naturalist 111: 1119–1144.

Cusack, D. and F. Montagnini. 2004. The role of native species plantations in recovery of understory woody diversity in degraded pasturelands of Costa Rica. Forest Ecol. Manag. 188: 1–15.

deGraaf, N.R. 1986. A silvicultural system for natural regeneration of tropical rain forest in Suriname: Ecology and Management of Tropical Rain Forests in Suriname 1. Wageningen Agricultural University, Wageningen, the Netherlands.

Denslow, J.S. 1980. Gap partitioning among tropical rain forest trees. Biotropica 12 (Suppl.): 47–55.

Dove, M.R. 1983. Theories of swidden agriculture, and the political economy of ignorance. Agroforest. Syst. 1: 85–99.

Dublin, H.T., A.R. Sinclair and J. McGlade. 1990. Elephants and fire as causes of multiple stable states in the serengeti-mara woodlands. J. Animal Ecol. 59: 1147–1164.

Dupuy, B., H.F. Maître and I. Amsallem. 1999. Techniques de gestion des écosystèmes forestiers tropicaux: état de l'art. Working Paper: FAO/FPIRS/05, Rome.

Egler, F.E. 1954. Vegetation science concepts. I. Initial floristic composition, a factor in old field vegetational development. Vegetatio 4: 412–417.

Fairhead, J. and M. Leach. 2001. Sustainable forestry in Trinidad? Natural forest management in the South-East. Working Paper from the project 'Forest Science and Forest Policy: Knowledge, Institutions and Policy Processes'. IDS, United Kingdom.

FAO. 1993. Ordenación y conservación de los bosques densos de América Tropical. Estudio FAO. Montes 101. Roma.

FDS/ITTO. 2012. Sustainable forest management and biodiversity conservation in Sarawak, Malaysia: following the 1989/1990 ITTO mission. Technical report by Forest Department Sarawak.

Fernandes, E.C.M. and P.K.R. Nair. 1986. An evaluation of the structure and function of tropical homegardens. Agricultural Systems 21: 179–310.

Gómez-Pompa, A. and F.W. Burley. 1991. The management of natural tropical forests. pp. 3–18. *In*: A. Gómez-Pompa, T.C. Whitmore and M. Hadley (eds.). Rain Forest Regeneration and Management. Man and the Biosphere Series, UNESCO and the Parthenon Publishing Group, Paris.

Government of Trinidad and Tobago. 2011. National Forest Policy.

Graham, R.T. and T.B. Jain. 1998. Silviculture's role in managing boreal forests. Conservation Ecology 2: 8.

Grandstaff, T. 1978. The development of swidden agriculture (shifting cultivation). Development and Change 9: 547–579.

Grebner, D.L., P. Bettinger and J. Siry. 2013. Introduction to Forestry and Natural Resources. Academic Press, LUGAR.

Groot, A., J.-M. Lussier, A.K. Mitchell and D.A. Maclsaac. 2005. A silvicultural systems perspective on changing Canadian forestry practices. Forestry Chronicle 81: 50–55.

Günter, S., B. Louman and V. Oyarzun. 2012. Criteria and indicators to improve the ability to monitor forests and promote sustainable forest management: interchange of ideas on the processes of Montreal and Latin America. Technical Series 54. CATIE. Turrialba, Costa Rica.

Haggar, J., K. Wightman and R. Fisher. 1997. The potential of plantations to foster woody regeneration within a deforested landscape in lowland Costa Rica. Forest Ecol. Manag. 99: 55–64.

Hagner, S. 1999. Forest management in temperate and boreal forests: current practices and the scope for implementing sustainable forest management. Working paper: FAO/FPIRS/03. FAO. Forest policy and planning division, Rome.

Hallé, F., R.A. Oldeman and P.B. Tomlinson. 1978. Tropical Trees and Forests: An Architectural Analysis. Springer-Verlag, Berlin.

Hallsworth, E.G. 1982. Socio-Economic Effects and Constrains in Tropical Forest Management. John Wiley and Sons, Chichester, United Kingdom.

Hendrison, J. and N.R. de Graaf. 2011. History of use and management of forests in Suriname. pp. 9–28. *In*: M.J.A. Werger (ed.). Sustainable Management of Tropical Rainforests: The CELOS Management System. Tropenbos International, Paramaribo, Suriname.

Hubbell, S.P. and R.B. Foster. 1992. Short-term dynamics of a neotropical forest: why ecological research matters to tropical conservation and management. Oikos 63: 48–61.

Ihalainen, L. 2007. Improved rubber agroforestry system Ras1 in West Kalimantan, Indonesia: Biodiversity and farmers' perceptions. M.Sc. Thesis. Department of Forest Ecology, Viikki Tropical Resources Institute, University of Helsinki.

Iskandar, H., L.K. Snook, T. Toma, K.G. MacDicken and M. Kanninen. 2006. A comparison of damage due to logging under different forms of resource access in East Kalimantan, Indonesia. Forest Ecol. Manag. 237: 83–93.

ITTO. 2002. ITTO Guidelines for the Restoration, Management and Rehabilitation of Degraded and Secondary Tropical Forests. ITTO Policy Development Series No. 13 in Collaboration with CIFOR, FAO, World Conservation Union (IUCN) & the World Wide Fund for Nature (WWF). Yokohama, Japan.

ITTO. 2003. Achieving the ITTO Objective 2000 and Sustainable Forest Management in Trinidad and Tobago: Executive Summary. The International Tropical Timber Council; Panama City, Panama.

ITTO/IUCN. 2005. Restoring Forest Landscapes: An Introduction to the Art and Science of Forest Landscape Restoration. ITTO Technical Series No. 23. Yokohama, Japan.

ITTO-FMB/DENR. 1995. Ten-year Production of Treated Residual Dipterocarp Forest Stands: Final Project Report in Collaboration with Forest Management Bureau -Department of Environment and Natural Resources-. Quezon City, Philippines.

Joshi, L., G. Wibawa, G. Vincent, D. Boutin, R. Akiefnawati, G. Manurung, M. van Noordwijk and S. Williams. 2002. Jungle rubber: a traditional agroforestry system under pressure. ICRAF. Bogor, Indonesia.

Kareiva, P., S. Watts, R. McDonald and T. Boucher. 2007. Domesticated nature: shaping landscapes and ecosystems for human welfare. Science 316: 1866–1869.

Klassen, A. 2002. Impediments to the adoption of Reduced Impact Logging in the Indonesian corporate sector. pp. 40–58. In: T. Enters, P.B. Durst, C. Applegate, P.C.S. Kho and G. Man (eds.). Applying Reduced Impact Logging to Advance Sustainable Forest Management. FAO, Regional Office for Asia and the Pacific. Bangkok, Thailand.

Lafont, P.B. 1959. The slash-and-burn (ray) agricultural system of the mountain populations of Central Vietnam. pp. 56–59. In: C. Ratanarat, C. Tongyai, W. Brink, N. Tongyai, D.V. Sassoon, C.F. Siffin and I. Farrar (eds.). Proceedings of the 9th Pacific Science Congress of Pacific Science Association, Bangkok, 1957, Vol. 7.

Lamb, D. and D. Gilmour. 2003. Rehabilitation and Restoration of Degraded Forests. IUCN, Gland, Switzerland and Cambridge (UK) and WWF, Gland, Switzerland.

Lamb, D., P.D. Erskine and J.A. Parrotta. 2005. Restoration of degraded tropical forest landscapes. Science 310: 1628–1632.

Lamprecht, H. 1989. Silviculture in the Tropics: Tropical Forest Ecosystems and their Tree Species—Possibilities and Methods for their Long-term Utilization-. GTZ: Eschborn, Germany.

Leslie, A.J. 1977. Where contradictory theory and practice co-exist. Unasylva 29: 2–17.

Lininger, K. 2011. Small-scale farming and shifting cultivation. pp. 89–94. In: D. Boucher, P. Elias, K. Lininger, C. May-Tobin, S. Roquemore and E. Saxon (eds.). The Root of the Problem: What's Driving Tropical Deforestation Today? Tropical Forest and Climate Initiative, Union of Concerned Scientists, Cambridge, UK.

Lowe, R.G. 1977. Experience with the tropical shelterwood system of regeneration in natural forest in Nigeria. For. Ecol. Manage. 1: 193–212.

MacDonald, G.E. 2004. Cogongrass (*Imperatacylindrica*) -biology, ecology, and management. Critical Reviews in Plant Science 23: 367–380.

Matthews, J.D. 1991. Silvicultural Systems. Oxford Science Publications, United Kingdom.

Mayer, H. 1980. Waldbau auf soziologisch-ökologischer grundlage, 2 Auflage. Gustav Fischer Verlag, Stuttgart-New York.

Mergen, F. and J.R. Vincent. 1987. Natural Management of Tropical Moist Forests: Silvicultural and Management Prospects of Sustained Utilization. Yale University, School of Forestry and Environmental Studies. New Haven, CT.

Monge, A. 2000. Evaluación de tratamientos silviculturales en tres bosques secundarios en la Región Huetar Norte de Costa Rica. COSEFORMA/GTZ/ITCR.

Nabuurs, G.J., O. Masera, K. Andrasko, P. Benitez-Ponce, R. Boer, M. Dutschke, E. Elsiddig, J. Ford-Robertson, P. Frumhoff, T. Karjalainen, O. Krankina, W.A. Kurz, M. Matsumoto, W. Oyhantcabal, N.H. Ravindranath, M.J. Sanz Sanchez and X. Zhang. 2007. Forestry. pp. 541–584. *In*: B. Metz, O.R. Davidson, P.R. Bosch, R. Dave and L.A. Meyer (eds.). Climate Change 2007: Mitigation. Contribution of Working Group III to the Fourth Assessment Report of the Intergovernmental Panel on Climate Change, Cambridge University Press, Cambridge, UK and New York, USA.

Nair, C.T.S. and C. Mammen. 1985. Forest management system in the tropical mixed forests of India. pp. 19–89. *In*: FAO. Review of Forest Management Systems of Tropical Asia: Case-Studies of Natural Forest Management for Timber production in India, Malaysia and the Philippines. FAO Forestry Paper 89, Rome.

National Research Council (Committee on Research Priorities in Tropical Biology). 1980. Research Priorities in Tropical Biology. National Academy of Sciences, Washington, DC.

Natural Resources Canada. 1995. Silvicultural Terms in Canada. Second Edition. Canadian Forestry Service, Ottawa.

Negi, S.S. 1996. Forests for Socio-Economic and Rural Development in India. MD Publications, New Delhi, India.

Negi, S.S. 2000. Himalayan Forests and Forestry. Indus Publishing Company, New Delhi, India.

Neil, P.E. 1981. Problems and Opportunities in Tropical Rain Forest Management. Occasional Paper No. 16, Commonwealth Forestry Institute. Oxford, UK.

Nielsen, U., O. Mertz and G. Noweg. 2006. The rationality of shifting cultivation systems: labour productivity revisited. Hum. Ecol. 34: 210–218.

Oldeman, R.A.A. 1990. Forests: Elements of Silvology. Springer-Verlag, Berlin.

Oldeman, R.A.A. and J. van Dijk. 1991. Diagnosis of the temperament of tropical rain forest trees. pp 21–89. *In*: A. Gómez-Pompa, T.C. Whitmore and M. Hadley (eds.). Rain forest regeneration and management. Man and Biosphere Series 6: UNESCO and the Parthenon Publishing Group, Paris.

Pelzer, K.J. 1948. Pioneer settlement in the Asiatic tropics. American Geographical Society: Special Publication No. 29. New York.

Popenoe, H. 1959. The influence of the shifting cultivation cycle on soil properties in Central America. pp. 72–77. *In*: C. Ratanarat, C. Tongyai, W. Brink, N. Tongyai, D.V. Sassoon, C.F. Siffin and I. Farrar (eds.). Proceedings of the 9th Pacific Science Congress of Pacific Science Association, Bangkok, 1957, Vol. 7.

Priyadi, H. and M. Kanninen. 2009. Towards Sustainable Forest Management by Implementation of Reduced-Impact Logging (RIL) Techniques in Indonesia with References in Kalimantan: Proceedings of Open Science Meeting. The Royal Netherlands Academy of Arts and Sciences (KNAW), Netherlands.

Redhahari, R., D.M. Lewis, P.D. Phillips and P.R. van Gardingen. 2002. A preliminary investigation into the effects of the thinning components of TPTI using the SYMFOR simulation model. University of Edinburgh, Edinburgh.

Rietbergen-McCracken, J., S. Maginnis and A. Sarre (eds.). 2007. The Forest Landscape Restoration Handbook. Earthscan, London, UK.

Ruthenberg, H. 1971. Farming Systems in the Tropics. Clarendon Press, Oxford, UK.

Savory, C.A.R. 1969. Crisis in Rhodesia. Oryx 10: 25–30.

Seydack, A.H.W. 2000. Theory and practice of yield regulation systems for sustainable management of tropical and subtropical moist natural forest. pp. 257–317. *In*: K. von Gadow, T. Pukkala and M. Tomé (eds.). Sustainable Forest Management. Vol. I in the book series Managing Forest Ecosystems. Kluwer Academic Publishers, Dordrecht.

Schlaepfer, R. and C. Elliott. 2000. Ecological and landscape considerations in forest management: the end of forestry? pp. 1–67. *In*: K. von Gadow, T. Pukkala and M. Tomé (eds.). Sustainable Forest Management, Vol. I. Kluwer Academic Publishers, Dordrecht.

Schmidt, R. 1987. Tropical rain forest management—a status report. Unasylva 156: 2–17.

Schütz, J.P., T. Pukkala, P.J. Donoso and K. von Gadow. 2012. Historical emergence and current application of CCF. pp. 1–28. *In*: T. Pukkala and K. von Gadow (eds.). Continuous Cover Forestry, Second Edition. Springer, Amsterdam.

Shugart, H.H. 1984. A Theory of Forest Dynamics: The Ecological Implications of Forest Succession Models. Springer, New York.

Similä, M. and K. Junninen. 2012. Ecological Restoration and Management in Boreal Forests—best practices from Finland. Metsähallitus Natural Heritage Services. Erweko Painotuote, Helsinki.

Sist, P., D. Sheil, K. Kartawinata and H. Priyadi. 2003. Reduced-impact logging in Indonesian Borneo: some results confirming the need for new silvicultural prescriptions. Forest Ecol. Manag. 179: 415–427.

Spencer, J. 1966. Shifting cultivation in Southeastern Asia. University of California: Geography 19. University of California Press, Berkeley.

Spiecker, H., J. Hansen, E. Klimo, J.P. Skovsgaard and K. von Teuffel. (eds.). 2004. Norway Spruce Conversion—Options and Consequences. European Forest Institute. Research report 18. Joensuu.

Spurr, S.H. and B.V. Barnes. 1980. Forest Ecology. Third Edition. John Wiley and Sons Inc., New York, NY.

Srivastava, P.B.L., A.M. Ahmad, K. Awang, A. Muktar, R.A. Kader, F. Che' Yom and L.S. See (eds.). 1982. Tropical Forests—Source of Energy through Optimisation and Diversification. Proceedings of the International Forestry Seminar. 11–15 November 1980. Universiti Pertanian Malaysia.

Stanturf, J.A. and P. Madsen. 2002. Restoration concepts for temperate and boreal forests of North America and Western Europe. Plant Biosystems 136: 143–158.

Swaine, M.D. and T.C. Whitmore. 1988. On the definition of ecological species groups in tropical rainforests. Vegetatio 75: 81–86.

Taylor, A.R., H.Y.H. Chen and L. Vandamme. 2009. A review of forest succession models and their suitability for forest management planning. Forest Science 55: 23–36.

Troup, R.S. 1928. Silvicultural Systems. Clarendon Press, Oxford.

Udell, R.W. 2003. Evolution of adaptive forest management in a historic Canadian forest. Paper presented at the XII World Forestry Congress, 21–28 September 2003, Quebec City. 0052-C1.

UNESCO. 1978. Tropical Forest Ecosystems. A State of Knowledge Report Prepared by UNESCO/UNEP/FAO. UNESCO-UNEP, Paris.

UNESCO. 2010. The Jarawa tribal reserve dossier: Cultural & biological diversities in the Andaman Islands. Edited by Pankaj Sekhsaria and Vishvajit Pandya. Unesco, Paris.

Vanclay, J.K. 1996. Lessons from the Queensland rainforests: steps towards sustainability. J. Sustain. Forest 3: 1–27.

Vega Peña, E.V. 2005. Algunos conceptos de ecología y sus vínculos con la restauración. pp. 147–158. *In*: O. Sánches, M.-H. Roberto, E. Vega, G. Postales, M. Valdez and D. Azuara (eds.). Temas Sobre Restauración Ecológica. Semarnat-INE. Mexico DF.

von Gadow, K. and J. Puumalainen. 2000. Scenario planning for sustainable forest management. pp. 319–356. *In*: K. von Gadow, T. Pukkala and M. Tomé (eds.). Sustainable Forest Management. Vol. I in the book series Managing Forest Ecosystems, Kluwer Academic Publishers, Dordrecht.

Waggener, T.R. 2001. Logging bans in Asia and the Pacific: an overview. pp. 1–42. *In*: P.B. Durst, T.R. Waggener, T. Enters and T.L. Cheng (eds.). Forests Out of Bounds: Impacts and Effectiveness of Logging Bans in Natural Forests in Asia-Pacific. FAO, Regional Office for Asia and the Pacific. Bangkok.

Watt, A.S. 1947. Pattern and process in the plant community. J. Ecol. 35: 1–22.

Webb, L.J. 1982. The human face in forest management. pp. 159–175. *In*: E.G. Hallsworth (ed.). Socio-Economic Effects and Constraints in Tropical Forest Management. John Wiley and Sons, Chichester, UK.

Weidelt, H.-J. 1996. Sustainable management of dipterocarp forests—opportunities and constraints. pp. 249–273. *In*: A. Schulte and D. Schöne (eds.). Dipterocarp Forest Ecosystems. World Scientific, Singapore.

Whited, T.L., J.I. Engels, R.C. Hoffman, H. Ibsen and W. Verstegen. 2005. Northern Europe: An Environmental History. ABC-Clio: Nature and Human Societies Series. Santa Barbara-Denver-Oxford.

Whitmore, T.C. 1984. Tropical Rain Forests of the Far East. Second Edition. Clarendon Press, Oxford, United Kingdom.

Whitmore, T.C. 1991. Tropical rain forest dynamics and its implications for management. pp. 67–89. *In*: A. Gómez-Pompa, T.C. Whitmore and M. Hadley (eds.). Rain Forest Regeneration and Management. Man and Biosphere Series 6: UNESCO and the Parthenon Publishing Group, Paris.

Wiersum, K.F. 1982. Tree gardening and taungya on Java: examples of agroforestry techniques in the humid tropics. Agroforest. Syst. 1: 53–70.

Wilkie, M.L., P. Holmgren and F. Castaneda. 2003. Sustainable forest management and the ecosystem approach: two concepts one goal. Working Paper FM 25, Forest resources division. FAO, Rome.

Yirdaw, E. and O. Luukkanen. 2003. Indigenous woody species diversity in *Eucalyptus globulus* Labill. ssp. *globulus* plantations in the Ethiopian highlands. Biodiversity and Conservation 12: 567–582.

Zagt, R. 2004. Issues Paper: Information issues in the Suriname forest sector. Tropenbos International Suriname Programme. Paramaribo, Suriname.

Zhang, Y. 2001. Institutions in forest management: special reference to China. pp. 353–364. *In*: M. Palo, J. Uusivuori and G. Mery (eds.). World Forests, Market and Policy. Kluwer Academic Publishers, Dordrecht, Netherlands.

CHAPTER 7

Challenges of Forest Conservation

J. Edgardo Arévalo[1,2,]* and *Richard J. Ladle*[3]

ABSTRACT

Forest conservation is entering a critical and challenging period due to a shifting mosaic of threats, policies, and actors. In this chapter we: (i) provide a brief summary of the history of forest conservation; (ii) identify the main threats to the preservation and biological integrity of the World's forests; and (iii) outline the key strategies that have been developed to address these challenges. We specifically highlight the development of macroeconomic mechanisms for forest conservation, many of which are a consequence of the recent alignment of forest conservation with the international agenda of climate change. Such approaches include market-based mechanisms such as the certification of forest products, payments for ecosystem services, carbon credits, clean development mechanisms, and REDD (Reduced Emissions from Deforestation and Degradation). Another promising financial tool for forest conservation is debt for nature swaps, which have enormous potential in much of the developing world. We conclude by identifying the key challenges for the conservation of forests in the 21st century, the most important of which will be to halt the continuing loss of habitat from the biodiverse tropical forests of the developing world.

[1] Center for Sustainable Development Studies, The School for Field Studies, Apdo.Post. 150-4013, Atenas, Alajuela, Costa Rica.
[2] Escuela de Biología, Universidad de Costa Rica, San Pedro, Costa Rica, ZC-2090
E-mail: jose.arevalohernandez@ucr.ac.cr
[3] Institute of Biological and Health Sciences, Federal University of Alagoas, PraçaAfrânio, Jorge, s/n, Prado, Maceió, AL, Brazil. CEP: 57010-020.
E-mail: richard.ladle@ouce.ox.ac.uk
* Corresponding author: earevalo@fieldstudies.org

Introduction

Forests are one of the most conspicuous ecosystems on the planet and have immense economic, cultural, and conservation importance. For example, they are home to numerous indigenous peoples, they protect watersheds, they store the largest stocks of carbon, they act as a source of invaluable food and firewood for the rural and low-income human populations, and they are home to vast amounts of biodiversity including some of the rarest species on the planet. Given the multiplicity of functions and roles, it is unsurprising that challenges of contemporary forest conservation are correspondingly numerous, complex, and dynamic.

In this chapter we: (i) provide a brief summary of the history of forest conservation; (ii) identify the main threats to the preservation and biological integrity of the World's forests; and (iii) outline the key strategies that have been developed to address these challenges.

History of Forest Conservation

Forests have always been at the forefront of nature conservation. One of the oldest protected areas in the world is the Main Ridge Reserve in Tobago, officially created on 13 April 1776 with the goal to protect the watershed that the colonial sugarcane plantations relied upon (Ramdial 1980). This is a remarkably early example of the use of protected areas for the creation of what is now known as ecosystem services (*sensu* Costanza et al. 1998) and illustrates that early conservationists already recognized many of the greatest challenges in forest conservation. Indeed, one of the main drivers of early conservation sentiment in the United States in the mid-19th century was the rapid demise of the once vast forests in the American great lakes region (Jepson and Ladle 2010). While widespread deforestation was invigorating a strong conservation sentiment in the US, elsewhere in the world efforts were directed to the rational and planned utilization of nature. The major proponents of these values, forerunners of the economic rationalism inherent in the ecosystem services concept, were the influential community of scientists supporting the vital forestry and agricultural sectors of the colonial endeavor—particularly Dutch scientists in Indonesia and British scientists in the Caribbean (Jepson and Whittaker 2002).

Another important historical element in the patchwork of values that infuse modern forest conservation is the idea that forests are intrinsic elements of the natural landscape with distinct cultural significance. This perspective was particularly developed by German foresters in the late 19th and early 20th centuries who became increasingly concerned that clear-felling policies were destroying magnificent specimen trees and despoiling areas of forests with special scientific and aesthetic value (Jepson and Whittaker 2002). Their response was to promote rational resource planning through the careful assessment and protection of culturally important attributes of nature. This, in turn, has led to the production of detailed forest and vegetation maps, the first versions of which were published in the first decades of the 20th century and which quickly became a standard tool for forest conservation and management.

By the early years of the 20th Century, it was almost universally accepted that the world's forests should be protected and managed, to maintain their valuable resources (mainly timber), sustain their important biophysical role in the water cycle, and because they are important elements of landscapes with unique cultural and aesthetic values. During the 20th century, these largely western-centered cultural values were considerably reinforced by the knowledge that vast stores of terrestrial biodiversity reside within tropical forests (mainly found in developing countries), and that these same forests were being destroyed at unprecedented rates leading to inflated levels of species extinctions. This knowledge soon became an essential element of modern conservation advocacy, propelled into the public sphere through high profile media campaigns by conservation NGOs and through popular science books. Possibly the most important example of the latter was the publication in 1979 of *The Sinking Ark: A new look at the problem of disappearing species* by the Oxford-based forester and conservation polemicist Norman Myers. It was in this book that Myers claimed (with very little scientific evidence) that as many as 40,000 species were going extinct every year due to tropical deforestation (Myers 1979).

The most recent chapter in the history of forest conservation has been the increasing realization that forests may have a pivotal role in mitigating the impacts of anthropogenic climate change. This has led to unprecedented levels of research and scientific cooperation to uncover the complex bi-directional interactions between the world's major forests and the regional climate. The recent alignment of forest conservation with the climate change agenda provides both potential opportunities and costs. With respect to the former, conservation of the world's forests has never had a higher profile, providing abundant opportunities for policy makers and practitioners to tap into diverse funding sources. The potential risks involved with such alignment are less obvious, but are nevertheless real. Foremost of these is the prioritizing of carbon benefits over biodiversity, potentially leading to further extinctions and the loss of biologically unique ecosystems. More generally, the explicit use of carbon benefits to target the protection of highly threatened ecosystems is yet to be fully developed.

Why Conserve Forests?

Forest conservation clearly has a long history and has been motivated by a wide variety of societal values. Initially, forests were protected to prevent indiscriminate exploitation of key natural resources such as timber or, less frequently, to maintain their role in the regional water cycle. Forests can also have important cultural significance and may be key constituents of the landscape, justifying the forest conservation because of their intrinsic qualities. More recently, the importance of forests as shrinking reservoirs for biodiversity and their potential role in mitigating climate change has made them a major target for contemporary conservation.

Global concerns about the alarming estimated rate of species loss in tropical forests was one of the key drivers of the landmark Convention on Biodiversity, signed at the Earth Summit in Rio de Janeiro in 1992. The Convention takes a clear utilitarian approach to conservation and developed a framework to ensure the protection of biodiversity. This approach has a clear focus on the sustainable use of natural resources and benefit-sharing from commercialization of genetic resources, and responds to the concerns of developing nations about the exploitation of genetic resources by western-based private institutions (Vale et al. 2008). There have been various knock-on effects of this perspective, perhaps the most critical of which is side-lining of older conservation values based on the intrinsic and cultural values of biodiversity (Ladle et al. 2011).

The last 30 years have also seen an increasing focus on the welfare and rights of indigenous forest communities and, to a lesser extent, of other communities and social groups that have historically exploited forest resources. Once again, of particular importance in tropical regions is that indigenous groups are becoming increasingly crucial actors in forest conservation policy and management. Most recently, the complex interactions between forests and global and regional climates (Bonan 2008) and the potentially devastating effects of climate change on forest structure (Nepstad et al. 2008) and biodiversity (Thomas et al. 2004) have reinvigorated debates about the most effective strategies for forest conservation.

In summary, conservation is moving towards a new era where the global focus is shifting from biodiversity towards the protection and management of ecosystem services such as the ability of forests to influence regional climatic conditions or their role as carbon sources or sinks. This policy shift is taking place during a uniquely dynamic period for many forests, driven by wide-scale changes in ecological assemblages and structure, weakening the resilience of these ecosystems to further environmental changes and, in the worst scenarios, pushing them towards ecological tipping points. One consequence of this shifting mosaic of threats, policies, and actors is that the conservation, management, and governance of forests has never been so complicated, or so challenging.

Deforestation: The Ongoing Threat

Despite historical and current efforts to preserve forests around the world, the challenges of effectively achieving forest protection goals are still numerous and complex. Although forest loss can also be attributed to natural disasters (e.g., fires), most of the enormous devastation of the Earth's forests over the last two centuries can be attributed to human activities. Since the rise of human civilizations, forests have been cleared to make room for growing crops, raising livestock, or to provide materials for building. Forest conversion to agricultural land began some nine to seven thousand years ago and has accelerated ever since but especially so during the last century. The continually increasing rate of forest loss is a consequence of both human population exponential growth and continually improving technology for forest conversion into other land uses. This process was responsible for the progressive depletion of the forests of the Middle East, the Mediterranean, Europe and, finally, the New World. Recent research indicates that about half of the original forest that once covered the terrestrial surface of Earth has been removed and that the war of attrition against forests continues at an unabated pace (Laurance 2010).

Forests can be either totally or partially removed to make room for other types of land use. While a clear-cut forest removal leaves little room for interpretation about the extent of loss, the degree of degradation caused by partially cleared areas (e.g., selective logging) varies greatly. One definition of forest degradation is the reduction of the capacity of a forest to provide goods and services (Lund 2009), stressing the utilitarian values of forests. However, other definitions incorporate parameters that include other values such as biological diversity, species composition, forest structure and function, soil composition, and carbon stocks (Simula 2009). Regardless of the degree of alteration of the species assemblage and biophysical structure of the original forest—from clearance to selective logging—human induced changes in forests are a leading cause of terrestrial biodiversity loss around the world (Whitmore and Sayer 1992; Brook et al. 2003; Riddle et al. 2011).

After enormous amounts of deforestation over the last three centuries or so, only about 39 million km^2 of forest remain at the 10 percent tree cover threshold (Schmitt et al. 2009). Despite the popular view that forest ecosystems are still disappearing, the extent of this loss and its associated consequences are still hotly debated (Laurance 2007). Such debates are partly driven by the difficulties of making accurate estimates of forest cover and degradation over large spatial scales. Indeed, accurate estimates of forest loss are difficult to obtain for a number of reasons: first, insufficient resources may limit many developing countries to produce reliable information about the status of their forests. Second, countries use different classification and assessment methods to survey forest cover and loss making comparisons and global statistics are difficult to obtain. The FAO (2010) reported an underestimation of 30,000 km^2 of forest loss between 1990 and 2000 after revising the figures using more precise information from cooperating countries. Moreover, differences in the parameters used to define "forest" can lead to significant discrepancies of forest loss estimates.

The calculation of net forest loss in different regions has also been problematic. Net forest loss is the balance between how much forest is lost (through deforestation) and how much forest is gained (through natural regeneration and reforestation) in a given period of time. Recent analyses have shown that while temperate forests in Europe and Asia have gained forest cover, other forests, such as tropical forests in South America and Africa, have experienced a comparatively large net loss (Fig. 1). The good

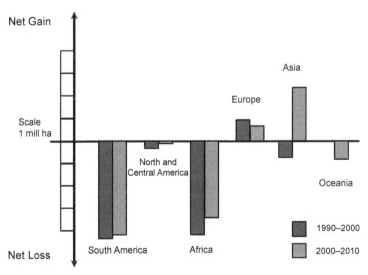

Fig. 1. Net forest loss and net forest gain in different regions of the world. Tropical forests of South America and Africa are greatly affected by deforestation whereas Europe and Asia are gaining forest through afforestation, defined as the conversion from other land uses into forests (based on FAO 2010).

news is that deforestation worldwide has dropped from 83,000 km^2 a year in the 1990s to 52,000 km^2 a year (an area roughly the size of Costa Rica) in the past decade.[1] However, such a coarse scale for global comparisons undoubtedly masks smaller scale variations of disproportionately high rates of forest destruction in areas such as Southeast Asia, Mesoamerica, and the Andes (Sodhi et al. 2004).

Tropical forests have been at the center of attention in terms of global conservation over the last 40 years. Such a focus is understandable, given that these forests maintain exceptionally high levels of biodiversity and, since recent decades, are being destroyed at a rapid rate. According to the Millennium Ecosystem Assessment report, about 9 million km^2 of tropical forests remain along the equator between the tropics of Cancer and Capricorn (N23.4° and S23.4°) (Duraiappah et al. 2005). Despite this convenient latitudinal delimitation, further complications arise when researchers try to reach a consistent and precise definition of "tropical forest". This is because there are numerous intermediate forms, along structural and biological continua, between sub-tropical and other woodland systems. Even when using the same forest definitions and similar methods, other studies assessing deforestation rates have yielded very different results. This suggests that factors such as variations in recording forest loss due to fires or different percentages of surveyed land can influence the final results of such surveys (see Lewis 2006).

Finer resolutions and precise estimates of net forest loss may help some tropical countries to tackle the conservation issue and to develop more effective strategies to prevent future biodiversity loss. Most importantly, the identification of areas in which at least 70% of their original vegetation has been lost is instrumental for the designation of high priority areas for biodiversity conservation. For example, this threshold is one component of the influential hotspots of the global conservation prioritization system (Myers et al. 2000).

Drivers of Forest Loss and Degradation

The factors driving forest loss are complex, interacting and operating across a variety of geographic and socio-political scales (Ladle et al. 2010). They include both ultimate factors such as subsidies, legislation, and international demand for forest and agricultural products, and proximate factors such as logging and the expansion of the agricultural frontier (Fig. 2).

---

[1] www.iucn.org/about/work/programmes/forest/iyf/facts_and_figure

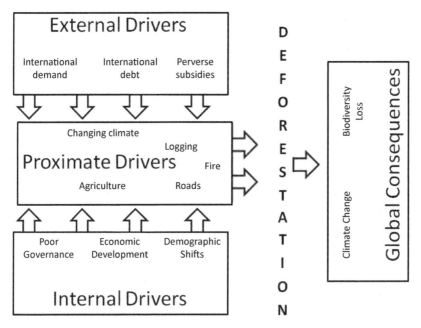

Fig. 2. Schematic representation of some of the major drivers of deforestation and their global consequences (based on Ladle et al. 2011).

Ultimate Factors

Temporal and spatial patterns of forest loss have historically varied greatly from region to region and more recently, from country to country. For example, human expansion in temperate regions such as the Mediterranean, Europe, and North America is thought to be one of the main reasons for large historical losses of different types of temperate forests (Kaplan et al. 2009). The underlying rationale is that population growth over a long period of time is associated to an ever increasing demand for food that, inevitably, leads to clearing of forests for agriculture. Another example, though at a much smaller scale, is the apparent relationship between increasing land clearing and population growth of the Mayan civilization in Northern Central America. The increasing population in the Copan area (650–900 AD) correlated with consumption of wood for fuel and construction as well as with land area for cultivation resulting in an increase in deforestation (Abrams and Rue 1988). Interestingly, recent research suggests that deforestation may have caused alterations in the regional climate contributing to frequent and severe droughts, which played a role in the eventual demise of the Mayan civilization (Oglesby et al. 2010).

Current estimates suggest that human population density is, or will become, a serious issue for some of the world's most biodiverse areas, including the remaining tropical lowland forests. Cincotta et al. (2000) estimated that in 1995 more than 1.1 billion people were living within the 25 biodiversity hotspots (Myers et al. 2000). Moreover, population growth rate within hotspots was estimated to be substantially higher than the global average, and even above that of developing countries. There is also good evidence that these increased human populations will have significant negative effects on biodiversity. For example, a study by McKee et al. (2003) found that human population density was strongly correlated with the number of threatened mammal and bird species across 114 continental nations, especially in those with large areas of tropical forests with high biological diversity.

Although the size of the human population may correlate with forest loss in many developing countries, population size per se is not the only ultimate factor driving environmental problems. Political organization, per capita income, insecure ownership, political turmoil, military unrest, international prices of goods and energy sources, and external debts are some of the interacting factors that can influence the rates of forest loss (Deacon 1994).

The distribution of human populations is also an important factor determining the degree of anthropogenic pressure on forests. Demographic shifts into forested regions are often heavily influenced by political policies, with a knock on effect for deforestation. For example, during the 1960s and 1970s many governments of developing countries implemented smallholder colonization policies under the U.S. program Alliance for Progress. These policies incentivized the colonization of pristine forested land by facilitating access through road infrastructure thereby preventing a costly agrarian distribution of the already owned land (Rudel et al. 2009). However, this process decreased in the 1990s when many governments began to preferentially incentivize highly capitalized enterprises to establish large cattle ranches and cultivate vast extensions of soybean and oil palm (Rudel et al. 2009).

Even if conservation policy is progressive and strictly implemented, the long-term success of forest conservation measures ultimately depends on public support. However, tapping into this support requires that the conservation message be carefully aligned with the target audience, whose expectations and values may differ between regions, countries, and to an extent, social groups. This is clearly illustrated in a recent study of media representations of Amazonian deforestation in the Brazilian and British print media (Ladle et al. 2010). In the U.K., the media highlighted the role of globalization in creating a demand for Brazilian agricultural products as the main proximate causes of Amazonian forest clearance. In contrast, the Brazilian news media generally showed far less interest in the external drivers of Amazonian deforestation and was much more tightly focused on internal drivers such as economic development and the effectiveness of national or local environmental policies.

It is unclear whether this pattern of representation is repeated in other developing and developed nations. Nevertheless, there has certainly been a strong tendency for western conservation organizations and conservationists to garner public support by highlighting the global significance of tropical deforestation for species extinctions or, more recently for ecosystem services. Indeed, NGOs such as Conservation International (CI) have been adept at utilizing a wide array of marketing strategies to get across their message about deforestation, extinction, and climate change. Perhaps surprisingly, this includes linking rainforest deforestation with American celebrity culture: in 2008 CI convinced the hirsute and globally famous actor Harrison Ford to have a patch of his chest hairs waxed (= ripped off) on camera. During this ordeal, he turns to the camera and declared that "every acre lost there [the rainforest of some developing world country], hurts here [in the developed world]!" (Jepson and Ladle 2010). However, it is unlikely that such celebrity-based media strategies would have the same impact on public opinion in the developed world, where conservation interventions need to be seen to balance biodiversity gains with economic and social progress.

In summary, the ultimate factors driving forest loss are complex and interacting, and include population growth, changing demographics, perverse incentives, and arguably, the need for economic growth driven by capitalist, free-market philosophies—although countries with different political ideologies have fared little better in protecting their forests. Reversing broader societal trends related to consumerism or eating behavior is clearly beyond the capabilities or remit of conservation. Nevertheless, the prospects for the long-term success and sustainability of forest conservation initiatives can potentially be improved by the careful alignment of the conservation message with societal values.

Proximate Factors

The immediate (proximate) causes of forest clearance are usually associated with the expansion of the agricultural frontier, logging, road building, and fires. These factors are typically interrelated, with road building allowing access to new areas of forest that are quickly exploited for valuable timber and cleared for small-holdings (Cropper et al. 1999; Perz et al. 2008). The influx of people into previously isolated areas is frequently associated with increased fire damage, either intentionally to clear forest for pasture/crops, or an unintentional consequence of human presence. Moreover, the knock-on effects of roads may be exacerbated by the construction of 'unofficial' roads that spring up near newly paved government-funded roads (Perz et al. 2008).

The effects of logging are often more difficult to assess since it is often selective and frequently illegal, making it difficult to monitor and to evaluate the environmental consequences. However, recent advances in remote sensing technology have started to reveal the enormous damage caused by loggers in some of the World's most valuable tropical forests. For example, a recent high-resolution remote sensing study in the top five timber-producing states of the Brazilian Amazon indicated that logged areas ranged from 12,075 to 19,823 km^2 per year ($\pm$14%) between 1999 and 2002, equivalent to 60 to 123% of previously reported deforestation area (Asner et al. 2005). Moreover, up to 1200 km^2 per year of logging were observed within protected areas, illustrating the enormous challenges of governing such activities in remote areas (Asner et al. 2005). Another study based on surveys of wood mills estimated that logging crews severely damage 10,000 to 15,000 km^2 per year of forests not included in deforestation mapping programmes (Nepstad et al. 1999).

Roads also have a major deforestation impact in terms of clearing new areas for agricultural expansion (Soares et al. 2004). In Brazilian Amazonia, the increase in deforestation since the 1960s was mainly caused by national policies supporting road building, tax and credit incentives to large corporations and ranches, and colonization projects for the rural poor. Smallholders in the Amazon have been estimated to clear between 2 to 3 ha per year per family in the first few years of settlement, but the rate is slowing down as crops (mainly rice, beans, maize, and cassava) are replaced by pasture (Fujisaka et al. 1996). Unfortunately, reducing deforestation by the rural poor has no easy solutions. Some models have suggested that policies seeking to intensify small-hold farming can somewhat reduce the rate of deforestation, although this result is dependent on both the distribution of land and the willingness of farmers to move to new areas (Rock 1996).

Roads also pose direct and indirect threats to wildlife populations as they may affect animal dispersal or cause mortality through vehicle-animal collisions (reviews in Forman and Alexander 1998; Forman et al. 2003; Coffin 2007). In this context, the expansion of roads and the associated increase in traffic volume would further exacerbate forest degradation though species loss. Recent research inside a protected area in Costa Rica strongly suggests the negative consequences of roads on biodiversity. A 4 km section of Carara National Park, located in the central pacific area of the country (Fig. 3), is delimited by a main highway and has experienced increasing traffic volume from 394.7 vehicles per hour in 2008 to 521.6 in

Fig. 3. View of Carara National Park delimited by the Coastal Highway (route 34), Central Pacific Conservation Area, Costa Rica (Photo credits: Roberto Ramos).

2012. This short stretch of road has reduced the diversity and density of birds in the surrounding forest (Arévalo and Knewhard 2011) and causes high mortality rates on thousands of vertebrates every year (J.E. Arévalo, unpublished data). Thus, although the actual forest area lost to roads might seem minimal, the various negative effects of roads contribute greatly to forest degradation (Forman et al. 2003).

Fire is often the main method to clear forest, both intentionally and unintentionally. Moreover, it is predicted that fires in tropical forests will increase as more damaged, less fire-resistant forests cover the landscape due to fragmentation, selective logging, and agricultural encroachment (Cochrane 2003). Indeed, some researchers argue that fire poses the greatest risk to tropical forest due to its interactions with other processes (Nepstad et al. 2001). For example, fires reduce precipitation by releasing smoke into the atmosphere. This, in turn, causes more fires. Likewise, fires kill trees, thereby increasing the susceptibility of forests to recurrent burning by opening gaps in the canopy and increasing material on the forest floor that can act as fuel (Nepstad et al. 2001).

Fires may also be related to climate change if a region experiences decreases in precipitation. Excess of atmospheric CO_2 (Lewis 2006) and nitrogen deposition from fires (Chen et al. 2010) can also affect forests by promoting plant growth (fertilization effect increasing NPP) or, in extreme cases, by causing dramatic shifts in ecology by tipping ecosystems over ecological 'tipping points' (e.g., Malhi et al. 2008). Conversely, forests themselves can have a substantial influence on regional and global climates because they account for ~40 percent of global NPP storing large amount of carbon (Cleveland et al. 2013). Thus, their removal could significantly alter climate regimes for remaining forests. Such effects may extend beyond the bounds of the biome, leading to strong synergies between deforestation in different regions. For example, Malhado et al. (2010) recently demonstrated that deforestation in the cerrado in central Brazil may be sufficient to significantly change the climate over some areas of the Amazon rainforest, in some cases crossing ecological thresholds of the biome putting the forests at risk of transformation into deciduous forest or even savannah vegetation. Forests also store a considerable amount of carbon, which is released during deforestation thereby contributing to global warming (Pan et al. 2011).

Current Approaches for Forest Conservation

The international conservation community has reacted to continuing tropical deforestation in a wide range of ways; most notably by establishing the Convention on Biological Diversity (CBD) in 1992. The CBD acknowledged that biodiversity is essential for human existence, but focusing strongly on the utilitarian value of biodiversity as a key to effective conservation. In other words, biodiversity must pay for itself, but only if the benefits arising from its use are fairly and equitably distributed (Gaston and Spicer 2009). The CBD obliges sovereign nations to develop national legislation, outlining strategies, plans and programmes that respond to the changing circumstances of biodiversity conservation in particular nations. Moreover, the CBD set ambitious global targets for biodiversity conservation. With respect to deforestation, the CBD has a target (reconfirmed in 2008), to effectively conserve at least 10% of each of the world's forest types.

The CBD may have set the framework within which nations develop their conservation strategies, but the cornerstone of global conservation action is undoubtedly protected areas (PAs) (Jepson et al. 2011). Globally, approximately 13.5 percent of all forests are under some form of protection, although this varies considerably across latitudes or continents (from 5.5% in the Palearctic to 13.4 percent in Australasia) and forest types (from 3.2% of temperate freshwater swamp forest to 28 percent of temperate broadleaf evergreen forest) (Schmitt et al. 2009). Nevertheless, this figure should not be taken at face value since many protected areas lack rigorous control leading to degradation of buffer zones, and frequently to logging within strictly protected areas (Curran et al. 2004). Moreover, tropical forests (especially those in South America and Southeast Asia) are becoming increasingly isolated, as the surrounding, unprotected forest is lost to agriculture, urban development, and fire (DeFries et al. 2005). More recently the role of PAs in maintaining regional climate dynamics has come under scrutiny. For example, a recent analysis by Walker et al. (2009) indicates that even if all the areas outside of PAs in the Amazon are deforested, this will not drive drier areas over their ecological tipping points. However, this may not be the case in other major forested areas that have considerably less protection than the Amazon, where approximately 54 percent of all forests are protected (Soares-Filho et al. 2010).

Restoration of Degraded Areas

Given that a large extent of the Earth's surface is already under some form of protection (see above), the scope for extensive increases in forest PAs may be limited. This firmly places the emphasis on reforestation and restoration of degraded areas as a way to maintain or even increase global forest cover, reunite ecologically isolated forest fragments, and meet diverse conservation objectives. Traditionally, forest restoration has relied on natural succession processes following agricultural or other land use abandonment; this accounts for the majority of forest regrowth in parts of Eastern Europe where population declines have led to wide-scale abandonment of cultivable lands (Baumann et al. 2012). It is also a potentially successful strategy for tropical forests, given enough time, protection from fire, and a viable seed-bank (Aide et al. 2000; Arroyo-Mora 2005). However, in the 1990s the focus of research and practice shifted towards ways to accelerate recovery, rapidly restore biodiversity, and increase productivity (Parrotta et al. 1997). One strategy is to use plantations to facilitate forest succession through modification of biophysical conditions in the regeneration area (Baumann et al. 2012)—this is often the only viable strategy when lands have been completely deforested with an accompanying change in soil chemistry and structure (Griscom and Ashton 2011).

One of the greatest challenges for forest regeneration is the sourcing of a wide diversity of native species—especially difficult in tropical ecosystems where tree diversity can be staggeringly high. Nevertheless, when appropriate seedlings are available the results of regeneration projects have been very promising. For example, in southwest Costa Rica, the restoration of 145 ha of abandoned pasture was stimulated by planting mixed stands of native hardwoods (Leopold et al. 2001). Planting native species to enrich degraded forest (enrichment planting) can also be an appropriate strategy, especially for low-diversity stands of early successional pioneer trees. Indeed, it has been estimated that planting deep-forest trees in pioneer stands may significantly reduce the impacts of fragmentation by attracting seed dispersers (Martinez-Garza and Howe 2003; Zahawi et al. 2013).

Market-based Mechanisms for Forest Conservation

Protection and restoration may still be the main strategies for forest conservation, but these are being increasingly complemented by so-called market-based mechanisms. The best known and arguably the most effective of these is the certification of forest products. Certification is achieved by creating a set of rules, categories, and criteria that provide consumers with a guarantee about the sustainability and traceability of a given forest product. The most well-known certification scheme is the one generated by the Forest Studies Council (FSC) for timber, which has had enormous success in persuading consumers to commercially boycott timber from non-sustainable sources. The underlying logic is that if the demand for non-certified timber is reduced, then the economic incentives for logging (legal and illegal) are also reduced.

Certification schemes can be effective, but it can be complicated and difficult to introduce and implement especially if the forest acts as a source for diverse products. When a faster response is required, it can be easier to negotiate directly with producers to modify their business practices. A recent example of this is the strategy that was adopted by conservationists to reduce the impact of palm oil plantations, which had expanded rapidly and were threatening some of the last remaining lowland forests in Sumatra (Fitzherbert et al. 2008). In 2001/2002, the international conservation organisation WWF initiated the Round Table on Sustainable Palm Oil (RSPO), involving big plantation companies and manufacturers of palm oil products.[2] The companies voluntarily agreed to practices that will reduce the impacts of palm oil plantations. However, the key to RSPO's success was a special scheme developed by the FSC known as High Conservation Value Forest (HCVF), which identifies forests within a plantation's forestry concession that are vital for conservation and should be protected.

The RSPO is an example of how market-based approaches are using corporate social responsibility as a tool for forest conservation. However, voluntary agreements are limited by the extent to which companies are willing to buy in to a given conservation scheme. This does not mean that conservation is powerless against non-compliant companies. One strategy is to pressure them by influencing companies further down the supply chain. A good example (described in detail in Jepson and Ladle 2010) is the Indonesian-owned company APRIL, which operates a huge pulp mill in

---

[2] www.rspo.org

Sumatra, which used timber from acacia plantations (in land converted from rainforest) and timber from more questionable sources. In 1997, Friends-of-the Earth (FoE) Groups in Finland and the UK discovered that APRIL owed large sums of money to eight leading financial institutions. These institutions were more sensitive to pressure coming from FoE than from ineffective regulations and quickly took remedial action to avoid becoming the target of a high profile conservation campaign. Shortly afterwards, WWF persuaded APRIL to adopt the HCVF scheme and tighten-up timber procurement and set aside a 100,000 ha forest block.

Perhaps the most ambitious market-based mechanism for forest conservation is the concept of paying for ecosystem/environmental services (PES) defined as "voluntary transactions where a well-defined environmental service is being bought by a buyer, if and only if the provider secures the provision of such service" (Wunder 2005). Forests provide various types of environmental services, but watershed protection and carbon storage have the highest profiles. Three broad types of PES are emerging: voluntary contractual agreements, public payment schemes, and the most ambitious of all, trading in environmental services.

Voluntary contractual arrangements are typically focused on relatively small geographic scales. An example of this is the arrangement between Inversiones La Manguera S.A. (a private company) and the Monterverde Conservation League (a non-profit organization described in detail in Bougherara et al. 2009). The company runs a 6 MW run-of-river power plant (*La Esperanza*) in northern Costa Rica and most of the watershed (3,000 ha) is located within the Children's Eternal Rainforest Reserve, owned by the NGO. Under a private agreement signed in October 1998 (Appendix 1 in Janzen 1999), *La Esperanza* pays the Monteverde Conservation League for environmental services (watershed protection) provided by their reserve. Although watershed protection is the key service for the efficient operation of the hydroelectric power plant, the conservation of the forest also contributes to mitigate climate change and to conserve biodiversity.

Public payment schemes are government led initiatives that require companies to pay a fee for ecosystems services, and then identifying areas on which to spend the generated income. One of the best examples is China's Forest Ecological Benefit Compensation Scheme, which involves levying an ecological tax on water and tourism businesses operating in scenic areas and then using this money to finance forest protection and restoration (Li et al. 2006).

The final mechanism, trading in environmental services, is both the most ambitious and the most problematic. Basically, a government or international organization sets a limit on the amount of ecological services that can be used. In the case of carbon trading, this limit is on how much carbon a company can emit. The government then issues quotas that can be traded. From the perspective of forest conservation, carbon trading is the focus of current action. The key concept in carbon trading is off-setting; because CO_2 has no geographic boundaries, companies and individuals who can't reduce their emissions can off-set them by buying into a new initiative that captures carbon or stops it from being released (Bumpus and Liverman 2008). This creates a market for 'credits' for tonnes of carbon emitted or sequestered (captured). The great hope is that it will create a financial incentive to avoid forests from being cut down, planting new ones, or restoring degraded ones.

There are two markets for carbon credits: the compliance market and the voluntary market. The cost of credits in the voluntary market is cheaper because they are not subject to the rigorous UN validation of credits traded under a flagship inter-government scheme called the Clean Development Mechanism (Streck 2004). A lot of effort is currently being expanded on efforts to bring forest conservation into the compliance market. The main mechanism for this is called REDD or Reduced Emissions from Deforestation and Degradation. Forest loss and damage contributes around 20 percent of all greenhouse gas emissions and REDD would enable countries to receive payments if they can reduce deforestation. Moreover, protecting or restoring forests based on their ability to store carbon will also protect biodiversity and potentially promote considerable synergies between these benefits (Phelps et al. 2012).

Despite significant progress towards becoming a universal and effective mechanism, REDD faces many challenges, such as: (i) how to know whether or not a forest would remain if the scheme was not enacted; (ii) how to set national and regional 'base-lines' of existing forest cover and deforestation rates against which carbon emission/sequestration rates can be measured (Olander et al. 2008); (iii) the possibility that, if a forest patch is protected, loggers or plantation companies might simply cut down another patch such that there is no net carbon gain—known as leakage (Dargusch et al. 2010); (iv) how to provide assurance that the forest is in the state it is claimed to be and remains in the claimed state; (v) how to quickly generate robust data on how much carbon different types of forest release when they are cut down, and how much carbon is sequestered by different types of forest growing under different conditions (Gibbs et al. 2007); (vi) how to ensure financial parity with destructive land-uses such as oil palm plantations (Butler et al. 2009); and (vii) how to factor in all of the socioeconomic costs of REDD, such as demographic shifts and declining tax revenues (Ghazoul et al. 2010).

The latest iteration of REDD, known as REDD+, takes into account the role of conservation, sustainable management of forests, and enhancement of forest carbon stocks. The ultimate success of REDD and REDD+ are difficult to predict. There seems to be a wide consensus among private and public actors that deforestation and climate must be addressed simultaneously (Pistorius 2012). Nevertheless, there is still an absence of reliable funding mechanisms. In addition, the complex reality of forest governance in developing countries is beginning to cause problems to the on-the-ground implementation. In conclusion, REDD and other mechanisms of trading environmental services offer the greatest hope for transforming the nature of global forest conservation, but there are still significant challenges to be overcome before the potential benefits for climate change, biodiversity, and sustainable development are realized.

Debt Swaps for Forest Conservation

First proposed by conservationist T. Lovejoy in 1984, the debt-for-nature swap has been frequently used as a financial tool to allow developing countries to achieve specific conservation goals (Cassimon et al. 2011). Through this mechanism, a developing country can both reduce its foreign debt and receive liquid funds to be invested in conservation projects. This mechanism was conceived by conservationists to reduce deforestation in developing countries that were heavily indebted and that also had low probabilities to comply with the payments to the foreign creditors (Sheikh 2010). In order for a debt-for-nature swap transaction to take place, an investor or donor needs to purchase the debt from the international creditor that agrees to sell a given amount for less than the face value and then redeem it in the debtor country at a face value of the local currency (Sheikh 2010). For example, Madagascar obtained an equivalent of US$1.7 million (in Malagasy francs) after Conservation International negotiated the face value of US$3.2 million debt for a purchase price of US$1.5 (Greiner and Lankester 2007).

This kind of debt swap is more likely to materialize in democratic countries and in tropical areas with a relatively high density of threatened species (Deacon and Murphy 1997). Such swaps may be of great benefit for conservation efforts in biodiversity priority areas. For example, debt-for-nature swaps generated $80 million to help the process of establishment of the conservation system in Costa Rica (Boza 1993; Greiner and Lankester 2007). Nevertheless, the scale and dimension of this financial instrument may not contribute significantly to the financial and conservation status of other countries (see Greiner and Lankester 2007; Cassimon et al. 2011). Shortcomings of the debt-for-nature swaps and political inconveniences of the involved parties have contributed to a decrease in the use of this financial instrument. However, the need for new alternatives to meet the international agenda towards schemes to reduce emissions from deforestation has led to the re-emergence of the debt-for-nature swaps (Cassimon et al. 2011). The scheme has been expanded to many other tropical countries through the Tropical Forest Conservation Act and it is estimated that some $218.4 million in local currency will be generated in the next 12–26 years (Sheikh 2010). Thus, the direct and indirect benefits, derived from the debt-for-nature swaps for the reduction of deforestation is highly dependent on the fair design of transactions, taking into consideration the context and idiosyncrasy of individual countries.

The value of debt-for-nature swaps to conservation organizations is arguably much greater than their actual financial value. For example, they attract positive media attention and show the participating organizations (often conservation NGOs) to be both entrepreneurial and economically sophisticated. Moreover, debt-for-nature swaps involve a powerful network that includes commercial banks and finance ministers, institutions and individuals who have previously been only peripherally involved in conservation. Finally, those conservation organizations involved in debt-for-nature swaps often play a major role in determining how funds are dispersed and what kind of conservation approaches are adopted (Jepson and Ladle 2010), helping to embed the conservation philosophies and perspectives into developing regions of the world.

Conclusions

Forest conservation is reaching a crucial period. Our planet is facing unprecedented environmental change and at the same time, the priorities of the global conservation movement are changing from the establishment of new protected areas to market based mechanisms and sustainable exploitation of forest resources. In this context, we highlight five main challenges for global forest conservation for the coming decade:

1. Slowing and eventually halting deforestation in hyperdiverse tropical forests of South America, Asia, and Africa,
2. Restoring ecological connectivity of fragmented forests in order to reduce species extinctions caused by small population size and to provide dispersal routes for species who's current range of distribution is becoming inhabitable due to climate change,
3. Developing effective, cost effective, and rapid to implement protocols for reforesting cleared areas, restoring structure, key ecological processes, and biodiversity,
4. Improving and expanding Payment for Ecosystem Services schemes, including REDD, whereby they become a key mechanism for conserving forests outside of protected areas,

5. Improving the governance, management, and enforcement of protected areas, preventing encroachment and exploitation of the most valuable forests that are already under some level of protection.

Meeting these challenges will require considerable investment and unprecedented cooperation between nation states, international organizations, the private sector, NGOs, and civil society. Whether these challenges are met forest conservation is likely to remain at the forefront of the global conservation agenda over the following decades. During the following century, we may easily witness a new pulse of mass extinctions driven by the destruction and fragmentation of the World's remaining tropical forests. Nevertheless, we consider that there is reasonable ground for optimism due to the rapid development of market-based mechanisms, and more broadly, because of the recent alignment of forest conservation with the global threat of climate change.

References

Abrams, E.M. and D.J. Rue. 1988. The causes and consequences of deforestation among the prehistoric Maya. Hum. Ecol. 16: 377–395.

Aguilar-Støen, M. and S.S. Dhillion. 2003. Implementation of the convention on biological diversity in Mesoamerica: environmental and developmental perspectives. Environ. Conserv. 30: 131–138.

Aide, T.M., J.K. Zimmerman, J.B. Pascarella, L. Rivera and H. Marcano-Vega. 2000. Forest regeneration in a chronosequence of tropical abandoned pastures: implications for restoration ecology. Restor. Ecol. 8: 328–338.

Arévalo, J.E. and K. Knewhard. 2011. Traffic noise affects bird species in a protected tropical forest. Rev. Biol. Trop. 59: 969–980.

Arroyo-Mora, J.P., G.A. Sanchez-Azofeifa, B. Rivard, J.C. Calvo and D.H. Janzen. 2005. Dynamics in landscape structure and composition for the Chorotega region, Costa Rica from 1960 to 2000. Agr. Ecosyst. Environ. 106: 27–39.

Asner, G.P., D.E. Knapp, E.N. Broadbent, P.J. Oliveira, M. Keller and J.N. Silva. 2005. Selective logging in the Brazilian Amazon. Science 310: 480–482.

Baumann, M., M. Ozdogan, T. Kuemmerle, K.J. Wendland, E. Esipova and V.C. Radeloff. 2012. Using the landsat record to detect forest-cover changes during and after the collapse of the Soviet Union in the temperate zone of European Russia. Remote Sens. Environ. 124: 174–184.

Bonan, G.B. 2008. Forests and climate change: forcings, feedbacks, and the climate benefits of forests. Science 320: 1444–1449.

Bougherara, D., G. Grolleau and N. Mzoughi. 2009. The 'make or buy' decision in private environmental transactions. Eur. J. Law Econ. 27: 79–99.

Boza, M.A. 1993. Conservation in action: past, present, and future of the national park system of Costa Rica. Conserv. Biol. 7: 239–247.

Brook, B.W., N.S. Sodhi and P.K. Ng. 2003. Catastrophic extinctions follow deforestation in Singapore. Nature 424: 420–426.

Brooks, T.M., M.I. Bakarr, T. Boucher, G.A. Da Fonseca, C. Hilton-Taylor, J.M. Hoekstra, T. Moritz, S. Olivieri, J. Parrish and R.L. Pressey. 2004. Coverage provided by the global protected-area system: is it enough? BioScience 54: 1081–1091.

Bumpus, A.G. and D.M. Liverman. 2008. Accumulation by decarbonization and the governance of carbon offsets. Econ. Geogr. 84: 127–155.

Butler, R.A., L.P. Koh and J. Ghazoul. 2009. REDD in the red: palm oil could undermine carbon payment schemes. Conserv. Lett. 2: 67–73.

Cassimon, D., M. Prowse and D. Essers. 2011. The pitfalls and potential of debt-for-nature swaps: a US-Indonesian case study. Global Environ. Change 21: 93–102.

Chen, Y., J.T. Randerson, G.R. van der Werf, D.C. Morton, M. Mu and P.S. Kasibhatla. 2010. Nitrogen deposition in tropical forests from savanna and deforestation fires. Glob. Change Biol. 16: 2024–2038.

Cincotta, R.P., J. Wisnewski and R. Engelman. 2000. Human population in the biodiversity hotspots. Nature 404: 990–992.

Cleveland, C.C., B.Z. Houlton, W.K. Smith, A.R. Marklein, S.C. Reed, W. Parton, S.J. Del Grosso and S.W. Running. 2013. Patterns of new versus recycled primary production in the terrestrial biosphere. P. Natl. Acad. Sci. USA 110: 12733–12737.

Cochrane, M.A. 2003. Fire science for rainforests. Nature 421: 913–919.

Coffin, A.W. 2007. From roadkill to road ecology: a review of the ecological effects of roads. J. Transp. Geogr. 15: 396–406.

Costanza, R., R. d'Arge, R. de Groot, S. Farber, M. Grasso, B. Hannon, K. Limburg, S. Naeem, R.V. O'Neill, J. Paruelo, R.G. Raskin, P. Sutton and M. van den Belt. 1998. The value of the world's ecosystem services and natural capital. Reprinted from Nature, pp. 253, 1997. Ecol. Econ. 25: 3–15.

Cropper, M., C. Griffiths and M. Mani. 1999. Roads, population pressures, and deforestation in Thailand, 1976–1989. Land Econ. 75: 58–73.

Curran, L.M., S.N. Trigg, A.K. McDonald, D. Astiani, Y.M. Hardiono, P. Siregar, I. Caniago and E. Kasischke. 2004. Lowland forest loss in protected areas of Indonesian Borneo. Science 303: 1000–1003.

Dargusch, P., K. Lawrence, J. Herbohn and Medrilzam. 2010. A small-scale forestry perspective on constraints to including REDD in international carbon markets. Small-Scale Forestry 9: 485–499.

Deacon, R.T. 1994. Deforestation and the rule of law in a cross-section of countries. Land Econ. 414–430.

Deacon, R.T. and P. Murphy. 1997. The structure of an environmental transaction: the debt-for-nature swap. Land Econ. 73: 1–24.

DeFries, R., A. Hansen, A.C. Newton and M.C. Hansen. 2005. Increasing isolation of protected areas in tropical forests over the past twenty years. Ecol. Appl. 15: 19–26.

Duraiappah, A.K., S. Naeem, T. Agardy and M.E. Assessment. 2005. Ecosystems and Human Well-Being: Biodiversity Synthesis. Island Press, Washington D.C., USA.

FAO. 2010. Global Forest Resources Assessment. Main report, Rome.

Fitzherbert, E.B., M.J. Struebig, A. Morel, F. Danielsen, C.A. Brühl, P.F. Donald and B. Phalan. 2008. How will oil palm expansion affect biodiversity? Trends Ecol. Evol. 23: 538–545.

Forman, R.T.T. and L.E. Alexander. 1998. Roads and their major ecological effects. Annu. Rev. Ecol. Syst. 29: 207–231.

Forman, R.T.T., D. Sperling, J. Bissonette, A. Clevenger, C. Cutshall, V. Dale, L. Fahrig, R. France, C. Goldman, K. Heanue, J. Jones, F. Swanson, T. Turrentine and T. Winter. 2003. Road Ecology: Science and Solutions. Island Press, Washington D.C., USA.

Fujisaka, S., W. Bell, N. Thomas, L. Hurtado and E. Crawford. 1996. Slash-and-burn agriculture, conversion to pasture, and deforestation in two Brazilian Amazon colonies. Agric. Ecosyst. Environ. 59: 115–130.

Gaston, K.J. and J.I. Spicer. 2009. Biodiversity: An Introduction. 2nd Edition. Wiley-Blackwell Publishing, 350 Main Street, Malden, MA, USA, 208pp.

Ghazoul, J., R.A. Butler, J. Mateo-Vega and L.P. Koh. 2010. REDD: a reckoning of environment and development implications. Trends Ecol. Evol. 25: 396–402.

Gibbs, H.K., S. Brown, J.O. Niles and J.A. Foley. 2007. Monitoring and estimating tropical forest carbon stocks: making REDD a reality. Environ. Res. Lett. 2: 045023.

Greiner, R. and A. Lankester. 2007. Supporting on-farm biodiversity conservation through debt-for-conservation swaps: concept and critique. Land Use Policy 24: 458–471.

Griscom, H.P. and M.S. Ashton. 2011. Restoration of dry tropical forests in Central America: a review of pattern and process. For. Ecol. Manage. 261: 1564–1579.

Guruswamy, L.D. 1999. The convention on biological diversity: exposing the flawed foundations. Environ. Conserv. 26: 79–82.

Janzen, D. 1999. Gardenification of tropical conserved wildlands: multitasking, multicroping, and multiusers. Proc. Natl. Acad. Sci. USA 96: 5987–5994.

Jepson, P. and R.J. Whittaker. 2002. Histories of protected areas: internationalisation of conservationist values and their adoption in the Netherlands Indies Indonesia. Environ. Hist. 8: 129–172.

Jepson, P. and R.J. Ladle. 2010. Conservation: A Beginner's Guide. One World, Oxford.

Jepson, P., R.J. Whittaker and S.A. Lourie. 2011. The Shaping of the Global Protected Area Estate. pp. 93–135. *In*: R.J. Ladle and R.J. Whittaker (eds.). Conservation Biogeography. Wiley-Blackwell Oxford.

Kaplan, J.O., K.M. Krumhardt and N. Zimmermann. 2009. The prehistoric and preindustrial deforestation of Europe. Quat. Sci. Rev. 28: 3016–3034.

Ladle, R.J. and P. Jepson. 2010. Origins, uses, and transformation of extinction rhetoric. Environment and Society: Advances in Research 1: 96–115.

Ladle, R.J., A.C.M. Malhado, P.A. Todd and A. Malhado. 2010. Perceptions of Amazonian deforestation in the British and Brazilian media. Acta Amazonica 40: 319–324.

Ladle, R.J., P. Jepson and L. Gillson. 2011. Social values and conservation biogeography. pp. 13–30. *In*: R.J. Ladle and R.J. Whittaker (eds.). Conservation Biogeography. Oxford University Press, Oxford.

Laurance, W.F. 2007. Have we overstated the tropical biodiversity crisis? Trends Ecol. Evol. 22: 65–70.

Laurance, W.F. 2010. Habitat destruction: death by a thousand cuts. pp. 73–88. *In*: N.S. Sodhi and P.R. Ehrlich (eds.). Conservation Biology for All. Oxford University Press, Oxford.

Leopold, A.C., R. Andrus, A. Finkeldey and D. Knowles. 2001. Attempting restoration of wet tropical forests in Costa Rica. For. Ecol. Manage. 142: 243–249.

Lewis, S.L. 2006. Tropical forests and the changing earth system. Philos. Trans. R. Soc. London 361: 195–210.

Li, W., F. Li, S. Li and M. Liu. 2006. The status and prospect of forest ecological benefit compensation. J. Nat. Resour. 21: 677–688.

Lund, H. 2009. What is a Degraded Forest. Gainesville, Forest Information Services, VA. USA.

Malhado, A.C.M., G.F. Pires and M.H. Costa. 2010. Cerrado conservation is essential to protect the Amazon rainforest. AMBIO 39: 580–584.

Malhi, Y., J.T. Roberts, R.A. Betts, T.J. Killeen, W.H. Li and C.A. Nobre. 2008. Climate change, deforestation, and the fate of the Amazon. Science 319: 169–172.

Martinez-Garza, C. and H.F. Howe. 2003. Restoring tropical diversity: beating the time tax on species loss. J. Appl. Ecol. 40: 423–429.

Matthews, E. 2001. Understanding the FRA 2000. Oceania 88: 129.

McKee, J.K., P.W. Sciulli, P.D. Fooce and T.A. Waite. 2003. Forecasting global biodiversity threats associated with human population growth. Biol. Conserv. 115: 161–164.

McNeely, J.A. 1999. The convention on biological diversity: a solid foundation for effective action. Environ. Conserv. 26: 250–251.

Muzaffar, S.B., M.A. Islam, D.S. Kabir, M.H. Khan, F.U. Ahmed, G.W. Chowdhury, M.A. Aziz, S. Chakma and I. Jahan. 2011. The endangered forests of Bangladesh: why the process of implementation of the convention on biological diversity is not working. Biol. Conserv. 20: 1587–1601.

Myers, N. 1979. The Sinking Ark: A New Look at the Problem of Disappearing Species. Pergamon Press, Oxford.

Myers, N., R.A. Mittermeier, C.G. Mittermeier, G.A. Da Fonseca and J. Kent. 2000. Biodiversity hotspots for conservation priorities. Nature 403: 853–858.

Nepstad, D., G. Carvalho, A.C. Barros, A. Alencar, J.P. Capobianco, J. Bishop, P. Moutinho, P. Lefebvre, U.L. Silva and E. Prins. 2001. Road paving, fire regime feedbacks, and the future of Amazon forests. For. Ecol. Manage. 154: 395–407.

Nepstad, D.C., C.M. Stickler, B.S. Filho and F. Merry. 2008. Interactions among Amazon land use, forests and climate: prospects for a near-term forest tipping point. Philos. Trans. R. Soc. London 363: 1737–1746.

Nepstad, D.C., A. Verissimo, A. Alencar, C. Nobre, E. Lima, P. Lefebvre, P. Schlesinger, C. Potter, P. Moutinho, E. Mendoza, M. Cochrane and V. Brooks. 1999. Large-scale impoverishment of Amazonian forests by logging and fire. Nature 398: 505–508.

Oglesby, R.J., T.L. Sever, W. Saturno, D.J. Erickson III and J. Srikishen. 2010. Collapse of the Maya: could deforestation have contributed? J. Geophys. Res. Atmos. 115: D12106.

Olander, L.P., H.K. Gibbs, M. Steininger, J.J. Swenson and B.C. Murray. 2008. Reference scenarios for deforestation and forest degradation in support of REDD: a review of data and methods. Environ. Res. Lett. 3: 025011.

Pan, Y., R.A. Birdsey, J. Fang, R. Houghton, P.E. Kauppi, W.A. Kurz, O.L. Phillips, A. Shvidenko, S.L. Lewis, J.G. Canadell, P. Ciais, R.B. Jackson, S.W. Pacala, A.D. McGuire, S. Piao, A. Rautiainen, S. Sitch and D. Hayes. 2011. A large and persistent carbon sink in the world's forests. Science 333: 988–993.

Parrotta, J.A., J.W. Turnbull and N. Jones. 1997. Introduction—catalyzing native forest regeneration on degraded tropical lands. For. Ecol. Manage. 99: 1–7.

Perz, S., S. Brilhante, F. Brown, M. Caldas, S. Ikeda, E. Mendoza, C. Overdevest, V. Reis, J.F. Reyes and D. Rojas. 2008. Road building, land use and climate change: prospects for environmental governance in the Amazon. Philos. Trans. R. Soc. London 363: 1889–1895.

Phelps, J., D.A. Friess and E.L. Webb. 2012. Win–win REDD+ approaches belie carbon–biodiversity trade-offs. Biol. Conserv. 154: 53–60.

Pistorius, T. 2012. From RED to REDD+: the evolution of a forest-based mitigation approach for developing countries. Curr. Opin. Environ. Sustainability 4: 638–645.

Ramdial, B.S. 1980. Forestry in Trinidad and Tobago. Forestry Division, Ministry of Agriculture, Land and Fisheries, Trinidad and Tobago.

Riddle, B.R., R.J. Ladle, S.A. Lourie and R.J. Whittaker. 2011. Basic biogeography: estimating biodiversity and mapping nature. Conserv. Biogeogr. 45–92.

Rock, M.T. 1996. The stork, the plow, rural social structure and tropical deforestation in poor countries? Ecol. Econ. 18: 113–131.

Rudel, T.K., R. Defries, G.P. Asner and W.F. Laurance. 2009. Changing drivers of deforestation and new opportunities for conservation. Conserv. Biol. 23: 1396–1405.

Schmitt, C.B., N.D. Burgess, L. Coad, A. Belokurov, C. Besançon, L. Boisrobert, A. Campbell, L. Fish, D. Gliddon and K. Humphries. 2009. Global analysis of the protection status of the world's forests. Biol. Conserv. 142: 2122–2130.

Sheikh, P.A. 2010. Debt-for-nature initiatives and the Tropical Forest Conservation Act: Status and implementation. Congressional Research Service, Library of Congress.

Simula, M. 2009. Towards defining forest degradation: comparative analysis of existing definitions. Forest Resources Assessment Working Paper, 154.

Soares, B., A. Alencar, D. Nepstad, G. Cerqueira, M.D.V. Diaz, S. Rivero, L. Solorzano and E. Voll. 2004. Simulating the response of land-cover changes to road paving and governance along a major Amazon highway: the Santarem-Cuiaba corridor. Global Change Biol. 10: 745–764.

Soares-Filho, B., P. Moutinho, D. Nepstad, A. Anderson, H. Rodrigues, R. Garcia, L. Dietzsch, F. Merry, M. Bowman and L. Hissa. 2010. Role of Brazilian Amazon protected areas in climate change mitigation. P. Natl. Acad. Sci. USA 107: 10821–10826.

Sodhi, N.S., L.P. Koh, B.W. Brook and P.K. Ng. 2004. Southeast Asian biodiversity: an impending disaster. Trends Ecol. Evol. 19: 654–660.

Streck, C. 2004. New partnerships in global environmental policy: the clean development mechanism. J. Environ. Develop. 13: 295–322.

Thomas, C.D., A. Cameron, R.E. Green, M. Bakkenes, L.J. Beaumont, Y.C. Collingham, B.F.N. Erasmus, M.F. de Siqueira, A. Grainger, L. Hannah, L. Hughes, B. Huntley, A.S. van Jaarsveld, G.F. Midgley, L. Miles, M.A. Ortega-Huerta, A. Townsend-Peterson, O.L. Phillips and S.E. Williams. 2004. Extinction risk from climate change. Nature 427: 145–148.

Vale, M.M., M.A. Alves and S.L. Pimm. 2008. Biopiracy: conservationists have to rebuild lost trust. Nature 453: 26–26.

Walker, R., N.J. Moore, E. Arima, S. Perz, C. Simmons, M. Caldas, D. Vergara and C. Bohrer. 2009. Protecting the Amazon with protected areas. PANAS 106: 10582–10586.

Whitmore, T.C. and J.A. Sayer. 1992. Tropical Deforestation and Species Extinction. Chapman and Hall, London, United Kingdom.

Wunder, S. 2005. Payments for environmental services: some nuts and bolts. CIFOR, Jakarta, Indonesia.

Zahawi, R.A., K.D. Holl, R.J. Cole and J.L. Reid. 2013. Testing applied nucleation as a strategy to facilitate tropical forest recovery. J. Appl. Ecol. 50: 88–96.

CHAPTER 8

Forests, Sustainability, and Progress

Safeguarding the Multiple Dimensions of Forests through Sustainable Practices

Robin R. Sears

ABSTRACT

In this chapter I review the linkages between forests and society, review a set of necessary conditions for the sustainable use and the conservation of forests, and then present a series of forest management practices that appear to be supportive of forest conservation. Of particular interest are the conditions for the applications of sustainable practices in different sectors in using, managing, and protecting forested landscapes. Three main messages emerge from the discussion: (1) a paradigm shift with regards to forests is underway at a conceptual level and global scale, but there is much work to be done to shift the thinking and practices at the national and local levels; (2) policies and institutions that are supportive of forest conservation, and particularly that are supportive of local stewardship of forests, are still quite weak; and (3) the economic case for forest conservation is not yet convincing. I examine the priorities and perspectives of distinct stakeholders and actors and the importance of forests for different sectors of society. While I consider multiple stakeholders, the details of this chapter are on the concerns of forest-

Center for International Forestry Research (CIFOR), Av. La Molina 1895, La Molina, Lima, Peru. E-mail: robin.sears@aya.yale.edu dependent communities with respect to the forest landscape and with respect to other actors, large and small. I describe some common forest conservation measures and forest management systems, their position in the paradigm shift, and their potential to contribute to productive forestry. I touch briefly on the enabling conditions, institutional structures and behaviors, and policy support that can facilitate forest conservation at the local scale. I conclude the chapter with a few points on ways forward.

Introduction

The premise of this chapter is that forests are worth conserving, and that there are effective and diverse ways of doing so. Forests and the biodiversity they harbor have immeasurable intrinsic value. They also serve as a fundamental component of sustainable rural livelihood strategies, providing both goods and services. Forests contribute significantly to national and subnational economies through the provision of timber resources, nature-based tourism, and in economic incentives such as REDD+ payment schemes for forest conservation for ecosystem services. Thus, given such recognition, this narrative has the potential to transform the way that human societies use and manage forests.

How we understand these multiple dimensions depends largely on one's perspective. We all depend on forests in some way, whether we know it or not. However, given the ongoing destruction and degradation of natural forests around the world, forests are not afforded the appreciation and protection they merit. For certain actors in the forested landscape, there is still too much profit in either selling off the rights to deforest, often through corrupt practices, unsustainable forest use, or deforestation. The problem is that many of those who degrade forests or deforest are not the ones who are directly impacted by their actions. In a new forest paradigm the multiple values of forests become clear and forest conservation, in its many guises, becomes the norm, and not the exception. Only when we will shift thinking away from reductionist tendencies—single use, single component—to holistic thinking about the forest as an ecosystem that is connected to social, economic, and political dimensions will we begin to manage our forests and safeguard the well-being of the people who depend on them.

In this chapter I use the term forest conservation broadly, referring to practices that result in the maintenance, management, or recuperation of forest cover, forest biodiversity, and forest function. The conservation value of a forest historically has been attributed largely to an appreciation for the intrinsic value of nature, of biodiversity. Today, population and economic growth can drive forest conversion to non-forest uses, but they also drive demand for more forest resources, particularly timber, fuelwood, and bush meat. People and industry are carving away at forest frontiers all over the world, and we allow it to happen at our own peril.

Nevertheless, it is nearly impossible to put a price tag on standing forest that will convince all of the actors to conserve forests. The market value of forest products and ecosystem services can be calculated and compared to the market value of alternative uses of that forest land, such as oil palm or timber plantation, or agricultural uses, but there are no coefficients for the social, cultural, and spiritual values of the standing forest, let alone the intrinsic value of biodiversity. Review Chapter 5 for details on measuring the value of forests. The argument to conserve forest cannot only be in economic terms; it requires an appreciation of the many dimensions of the value of forests.

In this chapter I review the linkages between forests and society in tropical forested countries, and consider approaches to forest management through the use, management, and protection of forested landscapes that integrate multiple values and uses of forests. Three main messages emerge from the discussion. First, a paradigm shift with regards to forests is underway at a conceptual level—from managing for single use to recognizing the multiple functions of forests, but there is much work to be done to shift the thinking and practices at the national and local levels. Second, despite ongoing and repeated forest reforms in many countries, policies and institutions that might support forest conservation, and especially local stewardship of forests, are still quite weak. Third, the economic case for forest conservation exists, but it is not yet convincing to the right people, namely, the ministers of finance and international agro-industrial investors.

While the priorities and perspectives of multiple stakeholders are considered, the details of this chapter are focused on the concerns of people who live in forested landscapes. Some forest conservation measures and forest management systems are described in the context of a forest paradigm shift. An attempt is made to identify the enabling conditions, institutional structures and behaviors, and policy support that can facilitate forest conservation at the local scale. I conclude the chapter with a few points on ways forward.

The Multiple Dimensions of Forest Use

Depending on geography and position in society, and on the knowledge and understanding of forests, different sectors and stakeholders view the roles and values of forests differently. At the local scale, forests are recognized as the basis for the well-being and livelihoods of people who are directly dependent on their goods and services. Forests are also appreciated for the extraction of natural resources—timber, minerals, oil, bush meat—that contributes to both local and industrial economies. Globally, forests of course serve in the regulation of carbon and water cycles, upon which our productive systems depend.

There may exist multiple and often conflicting aspirations for the same forest stand, often from competing forces both local and external to the region that drive land use decisions. For example, for a stand of mature forest in a remote district, the forestry office may grant timber concession rights to the stand, with the goal of generating income from logging concession fees. However, the timber concession may be superimposed on a community reserve that was designated by the environment office, and where local residents may rely on the forest for collection of medicines and plant fibers and subsistence hunting. And finally, another office in environment may anticipate the inclusion of that stand in the area in a new carbon accounting scheme that requires it to remain unaltered. Who gets to decide the fate of that forest? That is a question for political ecologists, who look at the power dynamics among actors.

As stated above, we are all dependent on the goods and services of forest ecosystems, whether we live in, at the margins of, or far from forests. But the degree of our dependency varies. A useful distinction to make is between people who rely on forest goods and services for their well-being and livelihood—and people whose only alternative for housing, for example, are the trees and palms in the forest, who rely on collected wood fuel, and whose only access to medicine may be the forest and their forest gardens. Others use forests and forest resources by choice: participants in forest-based extractive industries or tourism businesses that are based on non-consumptive use of forests, or multi-sited households who maintain ties to forest and farm lands. There are others who depend on the goods and services of forests, but who live far from their origin.

Among the forest-dependent people, Sunderlin and others (2005) in a paper examining the linkages between forests and rural poverty alleviation defined the basic typology of livelihood strategies of forest-dependent people as forest hunter/gatherer, to shifting cultivator in or at the margins of the forest, to sedentary agriculturalist at the forest frontier. Along this trajectory the actor's direct dependence on forest resources tends to decline as more food and income comes from agricultural crops and from off-farm labor. At the same time, due to over-exploitation and poor management, the availability of forest resources may decline (Sunderlin et al. 2005), leaving residents no choice but to look for substitutions or to make better use of the remaining resources. This often leads to migration to urban areas, but not always to the abandonment of their engagement with forests (Padoch et al. 2008).

Where they still exist, forests serve both as resource and as a matrix in which rural residents engage in a suite of livelihood strategies. Forests provide essential building materials for rural infrastructure: house construction materials, including wood for framing, walls and floors, palm fronds for thatch, and vines for binding come from forests. They provide myriad non-timber products, such as medicines, condiments, vines, bush meat, fungi, insects, birds; ecosystem services, such as indicators of seasonal or stochastic change, a role in the global carbon cycle; and aesthetic and spiritual values. The biodiversity of forests also house a vast genetic library that lends itself to natural evolution and can support the improvement of agricultural crops and development of medicines and pharmaceuticals. We need our forests.

Forests also play an important role as a matrix in which agriculture takes place, which may seem contradictory, especially when shifting cultivation is cited as a main source of deforestation in some parts of the humid tropics. An oft-cited example is that forests and other natural areas provide habitat for crop pollinators, an ecosystem service for agriculture with a value estimated at $194 billion annually (Gallai et al. 2009). In addition, maintenance of seed trees of economically important species around an agricultural field can yield fallow forests of high economic value. This subsidy from nature in the form of pollination services and seed source for natural regeneration of forests goes unrecognized by policy makers and other actors.

Even for those living far from the forest, we depend on the services forests provide. For example, the drought of 2014–15 in São Paulo state is linked to deforestation in the Amazon. Brazilian scientist Antonio Nobre and colleagues show that a single large Amazonian tree can evaporate up to 300 L of water per day, sending water vapor into the "flying rivers" that recharge aquifers and reservoirs far and wide. They show that "global warming and the deforestation of the Amazon are altering the climate in the region of Sao Paulo by drastically reducing the release of billions of liters of water by rainforest trees."[1]

While there is no definitive answer to the question of how much forest we need, and what kinds and in what configurations they should be, it is clear that the high fragmentation of forests worldwide presents a problem for biodiversity conservation and provision of ecosystem services (Haddad et al. 2015) and that forests are a crucial element in the balance

---

[1] http://in.reuters.com/article/2014/10/24/foundation-brazil-drought-idINKCN0ID1Y420141024

for a sustainable planet. The conceptual framework of "telecoupling" can be useful in defining the role of forests in life and livelihoods far and wide (Liu et al. 2013). This framework broadens the integrating concept of coupled human and natural dynamics to consider the socioeconomic and environmental interactions over distances and among multiple systems. Liu and colleagues (2013) illustrate the concept using the example of how the movement of invasive species around the world through trade has affected trade policies. In this telecoupled world, my forest practices may affect the price of your vegetables.

On the other hand, with increasing urbanization on the global scale, there is an apparent decoupling of many people's subsistence dependence on forests. Either by choice or necessity some portion of the rural population moves out of the forest, or away from the forest frontier, to towns and cities (Parry 2009), where the often better health services and educational institutions carry the promise of economic advance for the next generation. The move may be a short distance, it may be of short duration, and it may be circular, or seasonal, but it changes people's relationship to the land. Padoch and colleagues (2008) have highlighted the phenomenon of multi-sited households in Amazonia, whereby families with ties in both urban and rural areas maintain a relationship to the forest.

While one of the proximate causes of deforestation in some countries may be due to smallholder farm expansion, there are other significant direct causes, such as large-scale clearing for plantations of oil palm and timber and other agricultural commodities, which occur at a much larger scale and more permanently than smallholder farming. At the same time, there are a host of underlying driving forces of deforestation (Geist and Lambin 2002). The analytical framework of "telecoupling" is useful to see the near and distant indirect drivers of deforestation (Liu et al. 2013). Indeed, in an assessment of drivers of deforestation across 41 countries, DeFries et al. (2010) show a correlation between forest loss and two factors, urban growth and net agricultural trade per capita, both of which seem to increase the industrial-scale, export-oriented agriculture. They found that rural population growth was not a significant factor in forest loss, thus suggesting that decoupling rural livelihoods from forest resource dependence is not a sure bet for forest conservation. They conclude that these other forces are more significant in forest loss than small-scale frontier forest clearing and that efforts to reduce deforestation should focus on addressing these large-scale drivers.

Elements of the Paradigm Shift

It is evident that concerted and coordinated efforts at discovering, adapting, and employing effective practices in forest conservation are essential for the maintenance of forest ecosystem function and forest-dependent rural livelihoods. Forests and the forest sector are affected by global climate change impacts on local ecosystems and systems of land-based production, global demands for local resources, and international influence on local governance. Yet it is local communities, regional municipalities, and national entities that ultimately feel the greatest impacts of change and ultimately must govern the social and political processes related to natural resources to meet their own goals. It becomes evident that there is still plenty of work to do to make the case for forest conservation when a top level national forest authority in charge of balancing forest conservation and productive forestry asks for help to construct an argument about the value of standing forests versus alternative uses to convince her colleagues in the ministries of finance and production that supporting sustainable forest use is the right thing to do. But making the case is not simple.

What are the guiding principles for forest conservation today? Is sustainability still a useful concept and goal to guide forest conservation? Have we found evidence of the sustainability of any practice at all? Has forest loss and degradation been halted? The loss of biodiversity? The pollution of the air, soil, water? Several years ago Padoch and Sears (2005) argued for the conservation of the concept of sustainability as a management goal. Others now argue against it (Benson and Craig 2014), suggesting that the answers to these very questions is largely no. Has it worked? Padoch and Sears suggested that each person should discover and understand the problems that affect the health and well-being of our own and other communities and of the environments in which we live, and that we should understand our roles in creating those problems and be engaged collectively in solving them. In this way, through knowledge and understanding, we can discover how to sustain the future in ever more effective and inclusive ways. Benson and Craig suggest that we are beyond the point of hope of finding equilibrium in our engagement with nature, that sustainability is not only undefinable, but not possible. Rather, they suggest, the concept of resilience should now guide our efforts in managing ecosystems, in balancing socio-ecological systems. Whatever we call it, we've got to apply it to balance the needs of the people with the needs of the very nature that supports people.

Analytical and Planning Approaches

Because of the multiplicity of actors, interests, values and use of forests, which often conflict, it is widely recognized that approaches to forest conservation must consider the needs and capacities of multiple sectors, and at multiple scales. This can be part of a landscape approach, which captures the multiple components, both natural and human, and uses of a region. The principle of adaptive management in planning and implementation for rural development and conservation is taking hold, and particularly with a focus on resilience in socio-ecological systems.

Stakeholders and actors in different social, economic, and political sectors view and treat the forest differently. A conventional logger looks into a forest stand and sees money in the boles of the trees. In fact, members of the logging team see different things. A chain saw operator in Peru may see the shihuahuaco tree as the culprit that dulls his saw blade for the meager wage of $15/day. The logging boss may see in the same tree the $1000 he could get if he clandestinely sells the trunk to a cutrero tucked in a tributary around the bend, rather than the $250 he would earn on that trunk if he includes it in the bulk of logs going directly to the mill. In contrast, an ecologist views the forest as a complex system of flows of matter and energy, where the droppings of a bat colony that rests in a great shihuahuaco tree contain both the seeds and nutrients for the advanced regeneration of trees and shrubs in the forest. The oil palm industrialist sees the forest only as an obstacle to clear in the path of establishing vast extensions of plantation. A resident of the area—who hunts and fishes in this forest, who utilizes the vines and medicinal plants, who may have invited the logger to cut that tremendous shihuahuaco tree (to earn $30 for it), and who clears land at its margins—sees her family's survival in the forest.

Everything is connected, yet many things conflict. Applying the new forest paradigm across sectors should shift thinking away from reductionist tendencies—single use, single component—to holistic thinking about the forest as an ecosystem that is connected to social, economic, and political dimensions. Sustainability is at the core of the new forest paradigm, but this too is a contentious concept and elusive goal, as previously explained. But with this perspective, the sustainable forestry practitioner considers the next harvest cycle of trees from this forest. The conservationist considers the livelihood needs of resident people. The corrupt state agent ready to make the deal with the industrialist considers the opportunity costs of clearing mature rainforest. Learning from the landholders about the endogenous practice of applying the landscape approach can inform every aspect of society's relationship to the forest, from land use planning to rural development to conservation planning.

Because they live there, the forest-dwellers' perspective is akin to taking the landscape approach—considering and managing the area for an array of objectives, not just one. Temporary users and newcomers, as well as external influencers, such as the decision makers in the national ministries and subnational governments, may have a single goal: economic growth, or even zero deforestation at whatever cost. They may not consider the interconnectivity of the components of the ecosystem and of the linkages between society, economy and nature, but they must.

Development in forested landscapes is inevitable through urban expansion, agricultural expansion (particularly non-food commodities, such as oil palm and sugar cane for ethanol production), and high impact infrastructural development (such as roads, hydroelectric installations, and river ports). Even American states continue to lose forested areas to urban expansion (Nowak and Walton 2005). Adverse impacts of development on the environment, rural livelihoods, and climate systems can be mitigated by decision-making processes that apply the principles of the landscape approach (Box 1). These principles attempt to frame development goals and conservation goals in the same initiative, where necessarily there will be trade-offs, but where hidden synergies may emerge.

Box 1. Ten principles of a landscape approach.

These principles presented by Sayer and colleagues (2013) represent a consensus opinion on the topic of the landscape approach and have been adopted by the Subsidiary Body on Scientific, Technical and Technological Advice of the Convention on Biological Diversity. The application of these principles is not without challenges, principal among them problems with governance and poor institutional capacity. The forest sector requires not only conceptual changes, towards interdisciplinary systems thinking and inclusivity, but also changes in institutional structure, culture, and practice. Nevertheless, following these principles should help to lay a good foundation for initiatives that result in forest conservation while meeting the needs of society.

1. Continued learning and adaptive management
2. Common concern entry point
3. Multiple scales
4. Multifunctionality
5. Multiple stakeholders
6. Negotiated and transparent change logic
7. Clarification of rights and responsibilities
8. Participatory and user-friendly monitoring
9. Resilience
10. Strengthened stakeholder capacity

An important concept for the landscape approach for understanding relationships between people and forests is the production landscape.

Combining conceptual elements of landscape ecology, agroecology, and community ecology, the term production landscape refers to the aggregated mosaic of production areas and natural landscape elements and the ecological relationships among them. The term can be applied to landscapes in any biome, to systems under subsistence or industrial management, or no management at all, simply stewardship, and production refers to ecosystem services as well as goods. In rural areas the production landscape circumscribes the mosaic of fields, fallows, forests, and other natural and anthropogenic landscape elements, and is a useful and flexible unit of analysis in areas where production, protection, and extraction activities are practiced over space and time, and where distinct interactions (sometimes dependencies) between and among elements have been identified.

Related to the concept of the production landscape is the construct of the socio-ecological system (SES). The premise of SES is that today, in the Anthropocene, there are few, or even no natural systems without people, and no social systems without nature; that natural and social systems are inter-connected and co-evolving. The key to the functional socio-ecological system is resilience, or its "capacity … to absorb disturbance and reorganize while undergoing change so as to retain essentially the same function, structure, identity, and feedbacks"(Walker et al. 2004) (cited in Rist et al. 2014: 3). All three concepts—the SES, the productive landscape, and taking the landscape approach—can help to shift the forest paradigm from reductionist approaches to conservation to more holistic approaches of development and conservation.

The shift needs evidence, however there is rarely sufficient information that fully illuminates a problem or opportunity. Yet stakeholders and actors in the forest sector make decisions all the time in the context of insufficient data, information, and knowledge. One of the comments heard often in the policy arena is: "We don't have data to support that claim". From farmers who have sold their trees to millers who arrive to their farm gate with a portable saw we hear, "I don't know how many trees I sold. The logger offered me six thousand Soles for all the commercial trees in these two hectares of fallow forest, and I needed the money urgently, so I took the offer". From the National Forest and Wildlife Service lawyers who are working on the new regulations we hear, "We don't know how many people this will effect, but we know it is important", when discussing the potential impact of this new regulatory mechanism of granting timber permits to farmers on untitled land. Even from our own research team, "We don't have data yet that can give us an accurate estimate of the volume of bolaina that is being produced by smallholder farmers in the region."

We need data to generate information that improves our knowledge so that we can make informed decisions about forests. Research on forests, on forest users, on forest-dependent people, and on forest services and products must continue and should focus on the urgent and most relevant questions. Research capacity, both in terms of human and intellectual capital, and funding for research, is too often lacking in government agencies and even countries. An essential component of improving the forest sector in many countries is strengthening education and training programs in forestry, statistics, communication, public policy, and program management, among others. Support for education and research is essential for nations to have a cadre of well-trained scientists, forest managers, policy-makers, and resource users.

Operationalizing Innovative Approaches to Forest Management

Much remains to be done to conserve our forests while continuing our use of and appreciation for forest products, from fuel wood to furniture, and enjoying the subsidy from nature that are ecosystem services that forests provide. A fundamental step is to assess the values of forests (addressed in Chapter 5), which will be quite different for the disparate sectors of society. This is no easy task (May et al. 2013), given the multiple actors, sectors, and scales of interests in forest areas, not to mention the volatility and uncertainty of emerging markets for forest services, such as the carbon market or schemes for direct payments for other ecosystem services. While the recreational value of forests is sometimes considered, the cultural, social, and spiritual value of forests is usually ignored in valuation studies. Putting a price tag on the forest under alternative land use and market scenarios may be the only thing that speaks to the decisionmakers in the ministries of finance, production, or environment. "Help me construct an economic argument for the value of standing forest versus other uses", I heard recently from the director of the Peruvian National Forest and Wildlife Service, who has the minister of environment pressuring the forest service to halt deforestation, while the minister of finance implored her to develop policies and mechanisms that improve economic productivity of the forest sector.

An awareness and appreciation for the multiple values of forests and the need for sustainable forest management practices can be generated and promoted through research, education, and advocacy (Box 2). Formal pathways should be established and kept open between policy makers and development planners and the people who understand the realities of forests and forest management. A great deal also can be learned from the forest users, the practitioners (Sears and Steward 2012).

Box 2. Communicating research.

Another aspect of a forest paradigm shift is in science communication and in strengthening, and crossing, bridges between science and policy, research and development, and policy and implementation. The policy makers do not tend to read scientific papers or visit rural farms; and scientists rarely read the World Bank Poverty Reduction Strategy Papers or rural development plans from the local government office. Communication through digital media, short written briefs, workshops, and field visits can advance the understanding and discussions among actors.

A CIFOR research team in Peru was involved in the production of a video (MINAM 2014)[2] about the smallholder forestry systems we study. The motivation to make the video was to tell the story in an accessible medium about the existence of such production systems in Peru and to describe their multiple values. The video was broadcast several times on national television and posted on the Internet. The video featured three smallholder farmers who manage trees on their farms with various objectives, including the conservation of over-exploited timber species, creation of habitat for wildlife, and cyclical production of fast-growing timber in fallows for sale when a need for cash arises. We coupled the video launch with a written brief geared towards policy makers describing the forestry system as a justification for a set of policy recommendations that we hope shall encourage smallholders to produce more timber and shall allow them to engage in the market with fewer problems than they currently have. The result was that a wide diversity of people were able to learn from the landholders directly.

In this century, heightened awareness of the warming global climate and increasing evidence of its consequences lends certain urgency about forests. While pressures from the social systems, including population growth, human migration, inequities and corruption, and increasing global demand for food and forest products, are at the root of forest degradation and deforestation, all of this is happening in the increasingly urgent context of a warming global climate. In light of this urgency, it is worthwhile for policy makers and development planners to take stock of and learn from over thirty years of approaches to sustainable development (Blom et al. 2010). Comparative studies across countries and regions can also reveal opportunities and pitfalls not considered locally (Sills et al. 2014). Decades of failed conservation and development projects contrasted with a few successful initiatives have shown that a necessary condition for meeting conservation and sustainable forest management goals in productive landscapes (where people live) is local participation in both setting the agenda for research and project activities and in monitoring

---

[2] This video can be viewed at http://www.cifor.org/youtube/secrets-of-the-forest/, in Spanish with English subtitles.

and enforcement of rules (Howard et al. 1995; Hall 1997; Horowitz 1998; Castello 2004; Börner et al. 2013). Top-down efforts rarely work.

The paradigm shift in forestry requires an educational approach that integrates trans- and interdisciplinary concepts and practices into school and university curricula and learning practices that actively engage students with the topic, including field-based (Sears 2013) and topical expeditionary learning approaches (Rheingold 2012). It requires practices that are developed through implementor participation, be they forest engineers or forest farmers, and the training and empowerment of forest users and stewards to govern their resources from a position of power, not fear or weakness. It requires training foresters in the principles of sustainability, the practices of sustainable management and production, and in issues of gender and social equity and appreciation of cultural diversity.

Sustainable practice in relation to forests does not stop at the forest edge. Our consumer behavior, our career choices, our voluntary activities make a difference to forests, too, and our actions speak to policy makers, to forest stewards, and to our neighbors. To inspire action, reflect on your relationship with forests (Box 3).

Box 3. Give forests a face.

The forest sector is by nature contentious. A tree is a very large organism to remove from a forest stand, one that does not generally go unnoticed. Scaling up, the removal of an entire forest stand has tremendous visual and ecological impact. Wood products for construction, furniture, and fuel are visibly voluminous. It is impressive to watch a trainload of logs of boreal pine rumbling across southeastern Canada; truckloads of oak, hickory, or birch on the highways in the eastern U.S.; or barge after barge with giant heaps of shihuahuaco or lupuna logs one meter in diameter plying the rivers in Peru. These visuals elicit distinct feelings in different people. One may imagine the forests from which these trees have been extracted. What of the bats who depend on the shihuahuaco seed for food; where will the boreal foxes live; how has harvest of the oaks affected the wild mushroom habitat? Others may justly imagine the wood products these harvests will afford, or the employment of people along the timber value chain, from loggers and camp cooks to the urban workers in the furniture factory.

Reflect on your personal relationship with forests and forest products. Go outside, open your eyes and look for the tree or forest element in the landscape right where you are. How does the landscape make you feel? Does what you see of the tree or forest element give you gratification? Despair? Learning to read the landscape, in all its complexity, including the human and natural elements of it, can help people to make the connections among the needs, uses, sectors, and conflicts in that landscape.

Box 3. contd....

Box 3. contd.

Now find a forest product that is part of your daily life. The hardwood floor, a chair, chopsticks. What species of tree is it made of? Where and how was the tree harvested? Who gained, what was lost from the harvest? Can you answer any of these questions? If you had to pay a premium for having the wood product rather than a substitute, say, plastic chopsticks, would you?

Public Policy, Institutions, and Governance

Sustainable forest management is far from being only a technical challenge. Rather, it requires social and institutional capacity and support. Moving from awareness to action on sustainability requires public policy, rules, rights, and institutions that provide and support governance structures and mechanisms that allow for sustainable and equitable practices with regards to forests. The goal of all of these things should be to integrate the social, cultural, and ecological outcomes of forest use and conservation.

Research and experience shows that building effective institutional arrangements and establishing and enforcing property rights are key to good governance that has positive outcomes for the livelihoods of forest-dependent and forest-adjacent people and on the conditions of the forest resources (Andersson 2012; Guariguata and Brancalion 2014). It is also clear that when public policy is based on top-down governance, with sole attention to formal rules and regulations, and ignoring the local context (i.e., the biophysical, socioeconomic, social, cultural context), usually policy fails (Sears and Pinedo-Vasquez 2011; Andersson 2012). A study of the forest sector in Cameroon showed that "market failure, policy failure, institutional weakness, debt crises, and population growth are issues that intricately work together to produce the negative outcome that is being witnessed in the forest sector of Cameroon" (Mbatu 2009). This example is a clear signal for the need for multi-sectorial approach: health, economy, education.

At the same time, too much regulation can stifle engagement in the forestry sector: "evidence in agriculture and forestry shows that links between producers and markets are often weakened by bureaucratic politics and organizational processes" (Masipiqueña et al. 2008). These authors evoke the discussion initiated by Ostrom (1990) about the utility of fine-grained regulations versus coarse-grained. Where government and other governance institutions are weak, they have a low capacity to implement, monitor, and enforce the fine-grained regulations. This highlights the need to strengthen institutions, including local governance arrangements, along with getting the policy right (Box 4).

Property rights are at the core of many problems in the forest sector (Mendelsohn 1994; Larson and Ganga Ram 2012). Property rights include the set of socially defined institutions that govern the appropriation and use of natural resources. Clear recognition of the norms and agreements, both customary and *de jure*, should reveal who can benefit from forests, land, and other natural resources, and in what ways. Tenure rights may include a bundle of rights, including access, use (or withdrawal), management, exclusion, and alienation. When these rights are vague, contradictory, or not respected it becomes difficult for local people to utilize natural resources with any sense of security or to regulate and enforce the rules. Unclear property rights is especially problematic for forest management, and particularly investing in tree planting, given the long time-frame needed for eventual harvest of trees. Multi-sectorial policy, strong local governance, and institutions, supported by equitable property rights are all essential for fair and equitable forest use and conservation.

Box 4. Rooting forest policy in forest realities in Peru: a case study.

A usual approach governments take in addressing natural resource problems is through policy reform. The government of Peru started a sector policy reform in 2001 with a new forest and wildlife law (No. 27308) with an aim to encourage sustainable forest management through a large-scale concession system. By 2008 it was apparent that there were ongoing problems in the forest sector (Smith et al. 2006; Sears and Pinedo-Vasquez 2011). Deforestation and forest degradation were continuing at unacceptable rates, informality continued to rule the forestry sector, and the reputation of the forest authorities was low. With support from the international community, and, most importantly, with the political will for change, the government initiated a participatory process to revise the forest and wildlife law again, and to develop regulations that were based in information and experiences from a broad sector of society.

Indeed, specific attention has been paid to these actors in this new reform, and particularly to those who have maintained forest cover through the integration of agroforestry practices in their farming systems for decades on public lands. The director of the National Forest and Wildlife Service (SERFOR) reported that it was evident after the 2000–2001 law (Law No. 27308) and regulations were passed that small-scale coffee and cocoa farmers who have been farming for decades [on land categorized as forest and under the public domain] were left out; thus, a new mechanism was created to include them in the forestry sector through the contract for assignment of use for agroforestry systems (F. Muñoz Dodero, personal communication).

Our research group saw a window of opportunity to contribute to the policy reform process with evidence from research; we engaged in the participatory process with a specific focus on the opportunities for smallholder farmers in the new law (No. 29763) and the accompanying regulations. The group presented data and information that was generated through extensive field research on

Box 4. contd....

Box 4. contd.

smallholder forest management systems to multiple stakeholders in the forestry sector and participants in the policy reform process with the goal to root the new regulations relevant to this class of actors in the geographic, social, and economic realities encountered on the ground.

The team discussed specific recommendations directly with farmers and with policy-makers for creating a regulatory environment that supports the production and sale of timber in smallholder forestry systems (Sears et al. 2014). Specifically, they suggested that the new regulations should:

1. Describe clearly and explicitly the practices and places where timber is produced in the mosaic production systems of smallholder Amazonian farmers;
2. Create and mandate the implementation of a simple and clear process for smallholder farmers to register their timber production systems in a national registry and allow them to obtain permission to harvest, transform, and transport the timber; and
3. Establish a process and define government support for an office in the appropriate state agency of local and regional governments to operationalize the registry and monitoring systems, and to provide technical and informational support to small-scale timber producers.

The ultimate aim of their initiative to build a bridge from science to policy was to help shift the paradigm of forestry in Amazonia from timber extraction to production and to gain recognition and support for the important role smallholders play in the forestry sector in Peru. The team found that several factors in their approach helped to form and fortify the bridge necessary to cross the science-policy divide. First was for the research team to invest time to read, discuss, and understand the details of the new law and the proposed regulations. This allowed the team to engage productively in two-way discussions with the policy-makers. Second was to discuss the policy details with the farmer producers and other actors in the value chain to understand their position and response to the proposed regulations. Third, the team used video medium to present some details about the smallholder forestry system to the public and to actors in the policy reform process. Finally, the research team found that persistence, not insistence, was key to ensuring that essential, yet seemingly minor details did not slip through the cracks.

Getting the policy right is only one step in the process of change. The next step is to engage in the process of design and implementation of corresponding mechanisms and programs.

Forests in the Anthropocene

In the Anthropocene, the proposed geologic age of human impact on the Earth's systems (Lewis and Maslin 2015), we are seeing and will continue to see many changes to the Earth's systems. Driven by unsustainable use of natural resources, unexamined combustion of fossil fuels, and unjust generation and disposal of waste, the destruction and degradation of Earth's ecosystems, including forests, threatens their resilience and challenges our very dependence on nature. How do we envision society's relationship with forests in the next decade or century? Concerted and coordinated efforts at discovering, adapting, and employing practices in forest conservation are essential for the maintenance of forest ecosystem function, and to human well-being. While there is global concern about forests, it is local communities, regional municipalities, and national entities that feel the greatest impacts of change and ultimately must govern the social processes and natural resources to meet their own goals. Safeguarding the multiple functions of forests requires the landscape approach and multiple forms of forest management, including sustainable management of permanent production forests, multi-functional production landscapes, and restoration initiatives.

Permanent Production Forest

The natural forests of many timber-producing nations are largely under public domain, and the concession forestry system still dominates national forest strategies across the global south. Although there is some evidence of positive outcomes from timber concessions in natural forests in Africa and South America (Karsenty et al. 2008), the concession system has not proven to be the panacea that was hoped for in terms of ensuring sustained timber production in natural forests, of creating conditions for social and economic equity in the forestry sector, and for enabling government authorities to maintain control in the sector. The concession system in Indonesia has largely failed to meet production and conservation targets (McRae 1997; Barr 2002). The concession forestry system in Peru has thus far failed, even after a significant forest policy reform in 2000, to control illegal logging or to integrate the sector (Salo and Toivonen 2009; Sears and Pinedo-Vasquez 2011; Finer et al. 2014). Indeed, the sustainability of industrial logging in natural tropical humid forests is a highly contested topic (e.g., Putz et al. 2000; Keller et al. 2007; Putz et al. 2012; Zimmerman and Kormos 2012), one worth paying attention to, since a central premise of many forest policies in the humid tropics is that logging of mature natural forest can be sustainable. A systematic review of the literature on the balance of timber production and conservation in logged forests is recently underway and should provide important insights (Petrokofsky et al. 2015). Governments, foresters, and the forest industry continue to grant and participate in concession rights for logging in large, remote tracts of mature rainforest, forests classified as permanent production forests.

Relief from over-exploitation of timber in the humid tropical forests, however, does not seem to be on the horizon for many countries. Until timber resources in natural forests are economically depleted, the viability of production forestry (i.e., timber plantations, secondary forest management, farm-forestry) remains low. Contrary to popular notion, there is little to weak evidence that timber plantations reduce the pressure on natural forests (Ainembabazi and Angelsen 2014), one reason being that the timber species diversity and size in natural forests is far more attractive than timber growing in plantations. Where extensive forest cover still exists, such as in Amazonia, Central African Republic, and Indonesia, extractive logging from natural forests remains the cheapest source of commercial hardwood timber, even if it is illegal.

The Productive Landscape

Productive forestry activities are also undertaken in private and community landholdings. Because more and more parts of the world are human dominated landscapes, there is interest in the role that local tree and forest management systems play in the conservation of forests and provision of forest ecosystem goods and services (Dawson et al. 2013) on farms and in locally managed forests. There is much work to be done to assess the conservation value of farm-forest systems, indeed, and to increase the value of these systems. A considerable obstacle to promoting forest conservation in the human-dominated landscape, as in other local systems, is the lack of appreciation by policy makers for the value and potential of these systems not only to support rural livelihoods, but also to provide ecosystem services, even carbon sequestration, and agricultural and forestry products. Lacking in many countries, and regions of countries, are appropriate policy frameworks and institutions to support sustainable forest management in production landscapes.

A motivation for promoting the integrated ecosystem and landscape approaches to conservation that has been tested around the tropics is the supposition that improving productivity on farms, at the farm-forest interface, can help to take the pressure off the forests in protected areas (Ashley et al. 2006). This research team concluded from a study in multiple countries in sub-Saharan Africa that the effectiveness of agroforestry to alleviate pressure on protected area forests is limited due to "systemic market constraints, contradictions between development approaches and conservation objectives, and inconsistencies in institutional and regulatory frameworks" (p. 663)—and a weak understanding and appreciation of the values of agroforestry systems for both livelihoods and conservation (Ashley et al. 2006). Similarly, policies geared toward channeling forestry activities to sanctioned spaces, such as the 2001 concession forestry system in Peru, were expected to alleviate logging pressure from both community forests and forests in protected areas, but this has thus far shown to not be the case (Oliveira et al. 2007; Salo and Toivonen 2009; Finer et al. 2014).

What, then, are some alternatives? The conventional approach to addressing problems is to undertake policy reform, and to strengthen institutions for monitoring and enforcement of good rules, as discussed in an earlier section of this chapter. Another is to empower the forest dependent and forest-adjacent people to steward the forest.

The diversified production landscape of many farming systems comprises an array of spaces, including multi-storied, multi-species forest stands at different stages of development. The farmer utilizes both natural ecological processes and agricultural, silvicultural, and livestock management practices to produce an array of products and services on and around the farm, including trees and forests. In this farm-forestry system the spatial components with trees include house gardens, agroforestry spaces, fallows, and patches of mature forest. These trees and forests on the farm provide multiple ecological services and goods, such as supporting wild populations of animals, plants, and other taxa through the creation of habitat and provision of food (Balée and Gély 1989), as well as the provision of timber and non-timber forest products.

Smallholder farmers in Amazonia play an important role in the forestry sector in producing timber on their farms, in particular fast-growing species to supply local and national markets. Of particular interest on the farm is the agricultural fallow, or secondary forest. It is a dynamic component of production mosaic landscapes and is an essential ecological component to the farming system. Its fundamental purpose is to allow for the recuperation of soil fertility through natural regeneration and decomposition processes over several years, but it serves many others, including providing habitat for wildlife, carbon sequestration, and non-timber forest products (Bruun et al. 2009; Rerkasem et al. 2009; Marquardt et al. 2013). Establishing on clear, open spaces, fast-growing pioneer tree species are successful in the fallow. Thus, it presents an ideal niche for short-cycle timber production systems. Under these systems, smallholders facilitate the growth of timber species in their fields by managing natural regeneration of fast-growing timber species, as well as enriching their fields with desirable timber and non-timber species, including fruit species that attract wildlife.

The relationships at the farm-forest interface are multifaceted, sometimes apparently in conflict, but also with an element of dependency (Keleman et al. 2010). Good soil management helps the productivity in the crop fields, which can alleviate the need for the farmer to clear new land for cropping, i.e., deforest. Of course, good productivity yields income that the farmer could use to invest in labor to increase the cropping area, thereby resulting in more deforestation. On the other hand, the farmer recognizes the ecological advantages of maintaining forest cover on their landholdings and at the margins of their fields. The ecological services forests provide to farming systems include habitat for pollinators, barrier to germplasm pollution, attenuation of water flow from rainstorms and wind that can damage crops, and cooling the edge environment, among other things. The adjacent forest also provides the farmer with ready access to forest goods, such as vines for construction or carrying, forest fruits and other edible products, game animals, timber, and medicines, not to mention the cultural and/or spiritual benefits. Because they recognize the value—actually the farm subsidy from nature—of the services provided by forests and other components of the natural ecosystem, rural producers are invested in supporting the resilience of these local ecosystems.

Farmers use adaptive management and farming strategies to adjust to the changing biophysical landscape and market conditions. The diversity of spaces, products, and production and management practices on these farms lends an element of flexibility and adaptability to changes in the system. If the price of maize goes up, the farmer may dedicate more productive land to maize, perhaps harvesting the timber in an old fallow to make an additional field. Or if the papaya crop fails suddenly from a fungal attack, the farmer has other farm products on which they can rely. This on-farm diversity serves as a safeguard against shocks and allows farmers to respond to trends.

Forest Restoration

Increasingly, afforestation and reforestation through timber plantations, fallowing agricultural fields, and forest restoration on degraded soils are recognized as essential for forest conservation as well as for wood production. Plantation forestry is promoted at multiple scales for production of timber and promotion of forest cover, from extensive industrial plantations to farm-based mixed plantations. Forest restoration is an approach to conservation and production that has gained more attention recently with the emergence of REDD+ programs and mechanisms for benefit sharing for forest people. Restoration is the focus in regions where the original forest cover is long-since gone. In the Wet Tropics of Australia, for example, on the Atherton Tablelands of Far North Queensland, the ancient forests were degraded during the timber boom starting in the late 1800s, and when the timber industry went bust, those degraded forests were cleared for agriculture. Increasing recognition of the importance of forest on the Tablelands and a changing demographic has resulted in concerted efforts to reforest critical areas of the landscape for biodiversity conservation and provision of ecosystem services. Forest restoration is an expensive endeavor, however, in Australia's Wet Tropics ranging from USD5000 to 67,000 per hectare (Preece et al. 2013). Even in lowland Peru, tree planting for the restoration of degraded mining sites is estimated to cost between USD2000 and 3500 (Román 2014).

Rural farmers and forest-dependent people in Amazonia conserve over-exploited timber species through planting and management of natural regeneration in their field systems as well as in surrounding areas of disturbed post-logged forest (Summers et al. 2004; Putzel et al. 2013). This enrichment increases the ecological and commercial value of the landholding and provides income with the sale of individual or small numbers of these high-value timber trees.

On-farm enrichment is a viable alternative to state-sponsored reforestation programs, which are sometimes plagued by inadequate approach and lack of follow up. Farmers may have incentive to manage (plant, transplant, protect) high-graded species because they have a long-term relationship with the land on which they live and work, but there are multiple factors that influence their decisions. Summers et al. (2004) report from western Brazil that farmer engagement in tree planting tends to be positively influenced by factors such as secure land tenure, long-time residence on the landholding, and large size of landholding. On the other hand, sometimes simple stewardship motivates. Putzel et al. (2013) report from Peru that farmers engaged in this activity despite having insecure land tenure. One motivation seems to be common across the region, and that is farmers' past and present interest in investing in future forest resources for their children and grandchildren (MINAM 2014).

Conclusion

A new forestry paradigm integrates production and conservation of forests, issues, and concerns at the local to the global scale, and in a socially inclusive manner. Three main elements of the shift are, first, a consideration of the multiple dimensions of the problem and the solutions, including the social institutions, cultural practices, economic realities, and environmental conditions of the local in question. Second, attention to how the distribution of costs and benefits of any action or plan will affect different social groups based upon gender, economic class, ethnicity, and identity. And third, an understanding of how people's interactions with the natural environment are shaped by local institutions, rules and rights, and norms and practices. The landscape approach can provide a model for land use and resource management planning that can be implemented on any kind of landscape, even in urban and peri-urban areas, and mainstreamed at larger scales.

The most pervasive challenge in the forest sector is weak governance and poor institutional capacity. There is no technical fix for this, and the sector requires not only conceptual changes, towards systems thinking and inclusivity across gender, socioeconomic status, geography, among other divisions, but also changes in institutional structure, culture, and practice. For forest conservation to be effective, initiatives require a shift from project-oriented development planning towards a process orientation. This would necessarily refocus attention from efforts to produce short-term visible impacts in development projects, such as establishment of a tree seedling nursery for a reforestation project, to building institutional capacity and culture of planning, for example, through adaptive management planning processes. The three concepts of the socio-ecological system, the productive landscape, and taking the landscape approach can help to shift the forest paradigm from reductionist approaches to conservation to more holistic approaches of development and conservation.

Acknowledgements

In writing this chapter I have drawn heavily on my experiences in and research on smallholder farms in the Peruvian and Brazilian Amazon. Few of the concepts and recommendations are new, and I drew on a rich literature on forests and people from around the world, especially from South America and Southeast Asia. I settled on the central premise of the chapter after discussions with the executive director of the Peruvian National Forest and Wildlife Service (SERFOR), Fabiola Muñoz Dodero, whose mandate it is to carry out a participatory forest and wildlife policy reform, while also disrupting and rebuilding the institutional structure and culture of the forest service. The director faces just about all of the challenges mentioned in this chapter, and many others, but she does so with tremendous energy and innovation. My recent engagement with CIFOR has also provided the opportunity to contribute to what we hope is a paradigm shift in Peru, and thus, helps to ground the ideas I present here. The chapter was improved by astute suggestions provided by Sergio Molina.

References

Ainembabazi, J.H. and A. Angelsen. 2014. Do commercial forest plantations reduce pressure on natural forests? Evidence from forest policy reforms in Uganda. Forest Policy Econ. 40: 48–56.

Andersson, K. 2012. CIFOR's research on forest tenure and rights. CIFOR, Bogor, Indonesia.

Ashley, R., D. Russell and B. Swallow. 2006. The policy terrain in protected area landscapes: challenges for agroforestry in integrated landscape conservation. Biodivers. Conserv. 15: 663–689.

Balée, W. and A. Gély. 1989. Managed forest succession in Amazonia: the Ka'apor case. Advances in Economic Botany 7: 129–158.

Barr, C. 2002. Timber concession reform: questioning the "sustainable logging" paradigm. pp. 191–220. *In*: C.J.P. Colfer and I.A.P. Resosudarmo (eds.). Which Way Forward?: Forests, Policy and People in Indonesia. Resources for the Future, Washington D.C., USA.

Benson, M.H. and R.K. Craig. 2014. The end of sustainability. Soc. Natur. Resour. 27: 777–782.

Blom, B., T.C.H. Sunderland and D. Murdiyarso. 2010. Getting REDD to work locally: lessons learned from integrated conservation and development projects. Environ. Sci. Policy 13: 164–172.

Börner, J., S. Wunder, F. Reimer, R. Kim Bakkegaard, V. Viana, J. Tezza, T. Pinto, L. Lima and S. Marostica. 2013. Promoting Forest Stewardship in the Bolsa Floresta Programme: Local Livelihood Strategies and Preliminary Impacts. Rio de Janeiro, Brazil: Center for International Forestry Research (CIFOR). Manaus, Brazil: Fundação Amazonas Sustentável (FAS). Bonn, Germany: Zentrum für Entwicklungsforschung (ZEF), University of Bonn.

Bruun, T.B., A.D. Neergaard, D. Lawrence and A.D. Ziegler. 2009. Environmental consequences of the demise in swidden cultivation in Southeast Asia: carbon storage and soil quality. Hum. Ecol. 37: 375–388.

Castello, L. 2004. A method to count pirarucu: fishers, assessment, and management. N. Am. J. Fish. Manage. 24: 379–389.

Dawson, I.K., M.R. Guariguata, J. Loo, J.C. Weber, A. Lengkeek, D. Bush, J.P. Cornelius, L. Guarino, R. Kindt, C. Orwa, J. Russell and R.H. Jamnadass. 2013. What is the relevance of smallholders' agroforestry systems for conserving tropical tree species and genetic diversity in circa situm, *in situ* and *ex situ* settings?: a review. Biodivers. Conserv. 22: 301–324.

DeFries, R.S., T. Rudel, M. Uriarte and M. Hansen. 2010. Deforestation driven by urban population growth and agricultural trade in the twenty-first century. Nat. Geosci. 3: 178–181.

Finer, M., C.N. Jenkins, M.A. Blue Sky and J. Pine. 2014. Logging concessions enable illegal logging crisis in the Peruvian Amazon. Scientific Reports 4: 4719.

Gallai, N., J.M. Salles, J. Settele and B.E. Vaissière. 2009. Economic valuation of the vulnerability of world agriculture confronted with pollinator decline. Ecol. Econ. 68: 810–821.

Geist, H.J. and E.F. Lambin. 2002. Proximate causes and underlying driving forces of tropical deforestation. BioScience 52: 143–150.

Guariguata, M. and P. Brancalion. 2014. Current challenges and perspectives for governing forest restoration. Forests 5: 3022–3030.

Haddad, N.M., L.A. Brudvig, J. Clobert, K.F. Davies, A. Gonzalez, R.D. Holt, T.E. Lovejoy, J.O. Sexton, M.P. Austin, C.D. Collins, W.M. Cook, E.I. Damschen, R.M. Ewers, B.L. Foster, C.N. Jenkins, A.J. King, W.F. Laurance, D.J. Levey, C.R. Margules, B.A. Melbourne, A.O. Nicholls, J.L. Orrock, D.X. Song and J.R. Townshend. 2015. Habitat fragmentation and its lasting impact on Earth's ecosystems. Science Advances 1: e1500052.

Hall, A. 1997. Sustaining Amazonia: Grassroots Action for Productive Conservation. Manchester University Press, New York, USA.

Horowitz, L.S. 1998. Integrating indigenous resource management with wildlife conservation: a case study of Batang Ai National Park, Sarawak, Malaysia. Hum. Ecol. 26: 371–403.

Howard, W.J., J.M. Ayres, D. Lima-Ayres and G. Armstrong. 1995. Mamirauá: a case study of biodiversity conservation involving local people. Commonw. Forest Rev. 74: 76–79.

Karsenty, A., I.G. Drigo, M.G. Piketty and B. Singer. 2008. Regulating industrial forest concessions in Central Africa and South America. Forest Ecol. Manag. 256: 1498–1508.

Keleman, A., U.M. Goodale and K. Dooley. 2010. Conservation and the agricultural frontier: collapsing conceptual boundaries. J. Sustain. Forest 29: 539–559.

Keller, M., G.P. Asner, G. Blate, J. McGlocklin, F. Merry, M. Peña-Claros and J. Zweede. 2007. Timber production in selectively logged tropical forests in South America. Front. Ecol. Environ. 5: 213–216.

Larson, A.M. and D. Ganga Ram. 2012. Forest tenure reform: new resource rights for forest-based communities? Conservation and Society 10: 77–90.

Lewis, S.L. and M.A. Maslin. 2015. Defining the anthropocene. Nature 519: 171–180.

Liu, J., V. Hull, M. Batistella, R. DeFries, T. Dietz, F. Fu, T.W. Hertel, R.C. Izaurralde, E.F. Lambin, S. Li, L.A. Martinelli, W.J. McConnell, E.F. Moran, R. Naylor, Z. Ouyang, K.R. Polenske, A. Reenberg, G. de Miranda Rocha, C.S. Simmons, P.H. Verburg, P.M. Vitousek, F. Zhang and C. Zhu. 2013. Framing sustainability in a telecoupled world. Ecol. Soc. 18: Art. 26.

Marquardt, K., R. Milestad and R. Porro. 2013. Farmers' perspectives on vital soil-related ecosystem services in intensive swidden farming systems in the Peruvian Amazon. Hum. Ecol. 41: 139–151.

Masipiqueña, A.B., M.D. Masipiqueña and W.T. de Groot. 2008. Over-regulated and under-marketed: smallholders and the wood economy in Isabela, The Philippines. pp. 163–176. *In*: D.J. Snelder and R.D. Lasco (eds.). Smallholder Tree Growing for Rural Development and Environmental Services: Lessons from Asia. Springer, Netherlands.

May, P.H., B.S. Soares-Filho and J. Strand. 2013. How Much is the Amazon Worth? The State of Knowledge Concerning the Value of Preserving Amazon Rainforests. Policy Research Working Paper. World Bank, Washington D.C.

Mbatu, R.S. 2009. Forest policy analysis praxis: modeling the problem of forest loss in Cameroon. Forest Policy Econ. 11: 26–33.

McRae, M. 1997. Tropical forests: is 'good wood' bad for forests? Science 275: 1868–1869.

Mendelsohn, R. 1994. Property rights and tropical deforestation. Oxford Econ. Pap. 46: 750–756.

MINAM. 2014. Los Secretos del Bosque (The Secrets of the Forest). Page 0:24 AmbienTV. MINAM (Minsterio del Medio Ambiente), Government of Peru, Lima, Peru.http:// www.cifor.org/youtube/secrets-of-the-forest/

Muñoz Dodero, F., February 2015, personal communication.

Nowak, D.J. and J.T. Walton. 2005. Projected urban growth (2000–2050) and its estimated impact on the US forest resource. J. Forest. 103: 383–389.

Oliveira, P.J.C., G.P. Asner, D.E. Knapp, A. Almeyda, R. Galván-Gildemeister, S. Keene, R.F. Raybin and R.C. Smith. 2007. Land-use allocation protects the Peruvian Amazon. Science 317: 1233–1236.

Ostrom, E. 1990. Governing the Commons. Cambridge University Press, Cambridge, United Kingdom.

Padoch, C. and R.R. Sears. 2005. Conserving concepts: in praise of sustainability. Conserv. Biol. 19: 1–3.

Padoch, C., E. Brondizio, S. Costa, M. Pinedo-Vasquez, R.R. Sears and A. Siqueira. 2008. Urban forest and rural cities: multi-sited households, consumption patterns, and forest resources in Amazonia. Ecol. Soc. 13(2): 2.

Parry, L. 2009. Drivers of the rural exodus from Amazonian headwaters. Spatial changes in Amazonian non-timber resource use. Ph.D. Thesis. University of East Anglia, Norwich, United Kingdom.

Petrokofsky, G., P. Sist, L. Blanc, J.L. Doucet, B. Finegan, S. Gourlet-Fleury, J.R. Healey, B. Livoreil, R. Nasi, M. Peña-Claros, F.E. Putz and W. Zhou. 2015. Comparative effectiveness of silvicultural interventions for increasing timber production and sustaining conservation values in natural tropical production forests: a systematic review protocol. Environ. Evidence 4: 8.

Preece, N.D., P. Van Oosterzee and M.J. Lawes. 2013. Planting methods matter for cost-effective rainforest restoration. Ecol. Manage. Restor. 14: 63–66.

Putz, F.E., D.P. Dykstra and R. Heinrich. 2000. Why poor logging practices persist in the tropics. Conserv. Biol. 14: 951–956.

Putz, F.E., P.A. Zuidema, T. Synnott, M. Peña-Claros, M.A. Pinard, D. Sheil, J.K. Vanclay, P. Sist, S. Gourlet-Fleury, B. Griscom, J. Palmer and R. Zagt. 2012. Sustaining conservation values in selectively logged tropical forests: the attained and the attainable. Conserv. Lett. 5: 296–303.

Putzel, L., C. Padoch and A. Ricse. 2013. Putting back the trees: smallholder silvicultural enrichment of post-logged concession forest in Peruvian Amazonia. Small-scale Forestry 12: 421–436.

Rerkasem, K., D. Lawrence, C. Padoch, D. Schmidt-Vogt, A. Ziegler and T. Bruun. 2009. Consequences of swidden transitions for crop and fallow biodiversity in Southeast Asia. Hum. Ecol. 37: 347–360.

Rheingold, A. 2012. Unalienated recognition as a feature of democratic schooling. Democracy and Education 20: 1–8.

Rist, L., A. Felton, M. Nyström, M. Troell, R.A. Sponseller, J. Bengtsson, H. Österblom, R. Lindborg, P. Tidåker, D.G. Angeler, R. Milestad and J. Moen. 2014. Applying resilience thinking to production systems. Ecosphere 5: 1–11.

Román, F. 2014. Recuperación de áreas degradadas por minería en Madre de Dios. USAID, Lima, Peru.

Salo, M. and T. Toivonen. 2009. Tropical timber rush in Peruvian Amazonia: spatial allocation of forest concessions in an univentoried frontier. Environ. Manage. 44: 609–623.

Sayer, J.A., T.C.H. Sunderland, J. Ghazoul, J.L. Pfund, D. Sheil, E. Meijard, M. Venter, A.K. Boedhihartono, M. Day, C. García, C. Van Oosten and L.E. Buck. 2013. Ten principles for a landscape approach to reconciling agriculture, conservation, and other competing land uses. P. Natl. Acad. Sci. USA 110: 8349–8356.

Sears, R., P. Cronkleton, M. Perez-Ojeda del Arco, V. Robiglio, L. Putzel and J.P. Cornelius. 2014. Producción de madera en sistemas agroforestales de pequeños productores: Una justificación de política forestal a favor de los pobres en el Perú. Center for International Forestry Research (CIFOR), Bogor, Indonesia.

Sears, R.R. 2013. Unalienated Recognition at the Core of Meaningful Exchange Between School and Community. A Response to "Unalienated Recognition as a Feature of Democratic Schooling". Democracy and Education 21: Article 9.

Sears, R.R. and M. Pinedo-Vasquez. 2011. Forest policy reform and the organization of logging in Peruvian Amazonia. Dev. Change 42: 609–631.

Sears, R.R. and A.M. Steward. 2012. Education, ecology and poverty alleviation. pp. 17–37. *In*: F. DeClerck, J.C. Ingram and C. Rumbaitis del Rio (eds.). Integrating Ecology and Poverty Reduction: The Application of Ecology in Development Solutions. Springer-Verlag, New York.

Sills, E.O., S. Atmadja, C. de Sassi, A.E. Duchelle, D. Kweka, I.A.P. Resosudarmo and W.D. Sunderlin. 2014. REDD+ on the ground: a case book of subnational initiatives across the globe. Center for International Forestry Research (CIFOR), Bogor, Indonesia.

Smith, J., V. Colan, C. Sabogal and L. Snook. 2006. Why policy reforms fail to improve logging practices: the role of governance and norms in Peru. Forest Policy Econ. 8: 458–469.

Summers, P.M., J.O. Browder and M.A. Pedlowski. 2004. Tropical forest management and silvicultural practices by small farmers in the Brazilian Amazon: recent farm-level evidence from Rondonia. Forest Ecol. Manag. 192: 161–177.

Sunderlin, W.D., A. Angelsen, B. Belcher, P. Burgers, R. Nasi, L. Santoso and S. Wunder. 2005. Livelihoods, forests, and conservation in developing countries: an overview. World Development 33: 1383–1402.

Walker, B., C.S. Holling, S.R. Carpenter and A. Kinzig. 2004. Resilience, adaptability and transformability in social-ecological systems. Ecology and Society 9(2): 5.

Zimmerman, B.L. and C.F. Kormos. 2012. Prospects for sustainable logging in tropical forests. BioScience 62: 479–487.

Conclusions: The Survival of the Fittest

Carlos Rojas,[1] *Adam W. Rollins*[2] and
Sergio A. Molina-Murillo[3]

Throughout the pages of this book and many others, the odd historical relationship between our species and nature is portrayed within a matrix of inconsistencies from a long-term management perspective. We could argue that the achievement of conscious sustainable practices has only permeated a limited number of societies over time and that the majority have been in fact blinded by a type of anthropocentric chauvinism. The irony is that even in modern times, when we consider ourselves developed at the maximum, most of our conspecifics are still not raised within true sustainable systems.

Did we really get ourselves into the Tragedy of the Commons? Perhaps. After all, our species has shown and still demonstrates visible signs of obsessive behavior and denial. We have created a "technofixed" mentality and a "whack-a-mole" development strategy that have shaped the way we live in the XXI Century. With these new tools, we have been playing a game in which we whack problems and wait for the next ones to come up, but no real solutions are achieved. This system of interactions has also shaped the manner our species has perceived and tangled with forest resources. In spite of such a trial and error approach, for some societies we could have done and still can do better due to accumulated knowledge about our effect on the planet.

[1] Forest Resources Unit and Department of Agricultural Engineering, University of Costa Rica, San Pedro de Montes de Oca, 11501-Costa Rica.
E-mail: carlos.rojasalvarado@ucr.ac.cr
[2] Department of Biology and Cumberland Mountain Research Center, Lincoln Memorial University, Harrogate TN 37752, United States.
E-mail: Adam.Rollins@LMUnet.edu
[3] Department of Environmental Sciences, National University of Costa Rica, Heredia, 30101-Costa Rica; and Forest Resources Unit, University of Costa Rica, San Pedro de Montes de Oca, 11501-Costa Rica.
E-mail: sergiomolina@una.cr

The paradigm of the forest is thus, not a fixed archetype and this has never been the case. The manner in which human societies have interacted with forest resources has always been extremely contextual. Early scavenger and hunter-gather societies utilized natural resources with little to no thought of sustainability. As agriculture developed facilitating increased population sizes, the impacts on forest resources became considerable and exerted strong selective pressures on the forests with little to no consideration of long term practices. For a considerable time various societies simply considered their impact on the planet to be trivial as they believed that these resources were unlimited. More modern societies have exercised various levels of management and sustainable practices, but these have often been evaluated based on economic or political standards only. Furthermore, most practices and their associated datasets lack the temporal, spatial and even contextual scales needed to provide strong evidence to substantiate their claims. As evidenced throughout the book, human-forest interactions, albeit deep and intricate, have evolved.

With the birth of "ecological community thinking" scientists have sought to understand the natural processes that govern forest dynamics. Debates have risen and included positions from those who view forest communities as groups of species that function as "superorganisms" to the opposite end of the spectrum where communities represent groups living together by chance alone. The modern arguments focus on finding community assembly rules and are split between the so-called "niche hypothesis" and the "dispersal hypothesis". Under the former, the organisms involved must be coevolved relative to one another; whereas, the latter operates more as a null model where stochastic processes dominate. While these areas of research have produced a rich body of scientific literature and dialog related to sustainability, they have largely neglected the human dimension and socioeconomic realms. In fact a large portion of the ecological literature has studied and discussed forests as though they are independent from humans even though we have observed how our interactions modify the performance of natural systems. In fact, this book would probably have not included a chapter on climate change if it was written a few decades ago. It is precisely such flexibility yet intensity in our relationship with nature what makes the paradigm of the forest a complex system.

It is exactly the nature of such complexity that makes it difficult, if not impossible, to obtain general reproducibility among scientific studies and we may never be able to delineate universal laws that govern forest ecosystems. It does not matter how much we know, or how much experience we have accumulated. Every relationship between forests, individuals or societies, is unique and will not necessarily fit preconceived ideas. Moreover, a management strategy that works at one place under a specific set of conditions at a given time is not guaranteed to produce the same results in a different area or time. As such, each society must do their best to balance the individual, societal, economical, and governmental demands within the context of the evolutionary and ecological knowledge base available to those responsible for managing their natural resources.

In this manner, independent of our technological and academic capacity, individual or group perceptions of the world and the network of interactions shaping our relationship with the planet are very different. Each and every particular family of approaches deserves the dedication of time and effort to be understood, but socioeconomic differences across countries and a number of experiential inequalities make the latter as unfeasible as necessary for a planetary effort to take place. A number of social groups have understood this for a long time, in their own practice, and have created a myriad of possible means for sustainable living. Most others of course, have not yet reached that point. For this reason, respect and tolerance for different interpretations should delineate the path for a consensus during the present time. The idea at least, deserves to be seriously considered and not to be discarded as quickly as it has been in recent history.

One must consider the divides that may exist among citizens, resource managers, the government, and businesses. Each group has a different lens by which they view the world and by extension likely have a different view of sustainability. What exactly is each group trying to sustain? A lifestyle? A continual harvest of forest products and their associated income? Political boundaries? Transportation corridors? Aesthetics? Ecosystem services or something else completely? Of course each view of sustainability has some merit but how does one go about prioritizing theses needs and furthermore at what spatial and temporal scales should these be considered? This is undoubtedly a difficult and multifaceted question with no single correct answer. Yet this represents the reality of what societies must consider and ultimately act upon when utilizing their natural resources. There is, however, a larger context that needs to be considered, namely how do these decisions impact neighboring societies, countries, and ultimately the entire biosphere. This is truly a complex situation and one that deserves significant thought and discussion as explored throughout the chapters in this book.

The survival of the fittest race is more complex than just species outcompeting each other. Ultimately this situation extends beyond the basic Darwinian view of natural selection. Of course the principles of this classic idea are operating, but the forests are also experiencing some degree of unintentional artificial selection as a byproduct of human activities. But in turn, the changes imposed on the natural ecosystems undoubtedly then exert influence back on the human dimensions as well. Perhaps, in the context of human-forest interactions there is not really a true distinction between artificial and natural selection, unless one contends that humans are removed from the environmental context. Regardless, humans and their economic, ethical, social, and political constructs as well as the species involved in forest ecosystems are involved in battles for survival across several fronts.

This survival can be studied in each one of the levels made up by the complex system discussed above and throughout the book. In our life in this planet, as a consequence of our character as a biological species, there has always been fitter individuals and social groups for certain aspects of particular lifestyles. We know there has been human migrations and geographical establishments as a partial consequence of the latter. We have conquered the world that way. However, there have also been ideas, concepts and approaches that have prevailed for a period of time and have been essential for such conquering. In this manner, we can track the "survivors" of the interaction between our species and nature in a number of ways other than at a simple species-level. The term "fit" then, does not have to be restricted only to species but to the complete set of interactions among biological, social, economic, historical and cultural systems encompassing the life of our species in the planet.

As such, the "fittest" have always been contextual as well. What we have considered the best practices and the best approaches towards managing natural resources, forests included, have always been dependent on the particular conditions shaping the complex system itself. The latter is of course space and time-dependent. There is no standard answer to the ecological problems our societies face at an individual level, and thus, every case deserves study. That is the key to sustainability. The fact that some answers could be found to solve a problem somewhere else or at a different time in history provides clues for our own answers to be found but those solutions do not represent magic recipes. What we have read in this book, is that those stories of success have been the ones in which compromise and strong sense of inclusiveness are two of the "survivor" components shaping the system. These two characteristics are primarily developed through personal experience and are difficult to transfer to other people. In this manner, sustainability is something that should be lived and not only understood.

Understanding provides paths to deciphering complicated systems such as transport modes, infrastructure and software, classical engineering issues, but do not really solve experiential interactions between our species and natural resources. For the latter, it is necessary to comprehend the biosocial settings shaping our lifestyles. Comprehension is of course much more complicated than simple understanding. Complexity can make use of techniques developed to study complicated systems but the determination of the latter does not really make them suitable to fully study undetermined complex systems. In this way, economic, political, natural and social sciences are also necessary, besides technology-based disciplines, to provide an integrated approach for the achievement of consensual goals towards sustainability. We have seen quite a few examples of this type of approach in the pages of this book demonstrating that as a species, we can play different roles in the planet.

The notion that societies only develop through technological and scientific discoveries is risky in the sense that mental reductionism can be a byproduct of human specialization. This book has provided some insights on the topic. We have never seen as many specialists, kings and queens of their little section of the universe, as today. Some philosophers such as José Ortega y Gasset have even warned us about this type of monotone system generating what he has defined as "mass men" and "self-satisfied men". For this reason, an inclusive approach to managing our planet that also considers the disciplinary interaction mentioned earlier should be one logical step to test our fitter ideas and perceptions. In this moment in history, a celebration of the differences, disciplinary, cosmovisional and experiential among others, of every one of those "fit" personal universes is imperative to reach real massive sustainable practices. However, as easy as it sounds, the inclusion of all those elements in the generation of ideas to continue developing our history in the planet may simply be utopic due to the human nature. Independently on the latter, the comprehension of our own constraints is exactly what can put us on the right tracks for feasible potential achievements.

Our planet has never seen the imbalance in resource reallocation that we observe today. We often refer to this as "contamination", "degradation" and "depletion", but what these definitions really explain is the manner in which we interact with our surrounding systems. As we have read in this book, we can track a good portion of changes in the interactive paradigm with forest resources back to the industrial revolution. Since then, and with the development of special machinery and anthropogenic production strategies we have increased the velocity of interactive changes with natural and forest resources. We have also increased the human population and boosted the consumption of energy needed by our species to "survive" according to a new archetype of what human life should be like. This type of "mass man" society has been controversial during its existence but still provided the means for most individuals to find themselves fit in their little section of the universe. After all, we cannot forget that human individuals compete with each other even from a genetic basis.

Given the wide misperception that humans are disconnected from the general environment and natural ecosystems poses serious challenges when trying to develop management strategies that seek to sustain biodiversity and protect ecosystem services while adequately balancing social, political, and economic demands. Part of the answer may lie in educating the various stakeholders about the different issues in a way that allows them to gain an appreciation that they are not independent from the rest of the world and that their actions impact the planet in ways that influence the evolution of both human dimensions and natural environments at all scales. This could readily be accomplished through examining the reported impacts of human activities on the availability of potable water, erosion, nutrient cycling, and global climate change as well as many other examples.

What we need to remember is that whatever our unique relationship with nature is, it still represents a personal form of survival. In our own reality, such "survival of the fittest" is an active process that changes over time just like our personal "paradigm of the forest" does. That is the natural process that should continue re-shaping both. Our species has evolved in an environment that provided elements for individual, and as a byproduct, group survival. Few macro-contextual elements have changed since our species diverged from other lineages, and this makes the resistance to conceptual change highly unnatural. Our relationship with the forest has always been unique and it will continue to be that way. We should not forget that our lifestyle is very much dependent on natural resources and as a consequence, at least try to interact less aggressively with the planet. Not only our personal, but our species survival and our direct lineages depend on that.

Nature Appropriation

Social scientists tell us today that most modern societies desire to strengthen their level of proximity with nature. Somehow we have forgotten, at least consciously, that our species evolved in nature and that our brains are still constructed for interactions with forest environments. Ironically, most of our economic and social systems are still nature-based and heavily rely on the resilience of environmental systems. However, with the degradation of the latter, we have limited the capacity for natural systems to cope with extreme events and with that, our own capacity to sustain the prevalent socioeconomic structure. In the last decades we have seen the devastating consequences of natural events such as earthquakes, tsunamis and floods on human population centers. Besides the thousands of human lives affected by those events, the economic toll has been impressively high as well.

Immersed in the explained system of development, most of us do not even think that our obsessive need for bringing "greenness" into urban centers may be a byproduct of our long-term relationship with natural resources. After all, in the evolutionary path of our species, it has taken only a few decades, a couple of centuries maybe, to dramatically change our paradigm of interaction with nature. Our deep evolutionary roots within natural systems, product of thousands of years of modifications, still shape for the most part the manner we interact with the planet. Several cognitive elements necessary for group survival such as social behavior regulation and observational learning have deeply evolved within a system of human-nature interactions. In this sense, simple aspects of life such as drinking water are hard to trigger a mass reaction towards a rational utilization of common goods in spite the fact that water is a key resource for personal and group survival.

The need for bringing nature into our lives goes largely unrecognized but unconsciously imperative. We have created urban settings and centered our lifestyle on a routine around the concepts of "work" and "income" but still promote vacation time out in the woods, at the beach or in any type of system that stimulates interaction with natural resources. Interestingly though, while these connections are massively sought, a large number of people in many societies do not actually interact with the natural environment. They simply view nature by road-side pull-off spots or simply through the windows of their cars and rental houses.

We have brought plants and animals into our homes and work places and these days we are talking about carbon offsetting our activities and promoting political compromise with global environmental issues. The latter may be more complicated than bringing a plant home but still demonstrates that unconsciously we show a level of nature appropriation and desire of control over it. In spite of that, we may still deny that the opposite, natural control over our species, is also very well established in the form of epidemics, population control-related genes, genetic drift and other forms of evolutionary pressure.

One of the issues with our modern lifestyle is that in a world of "self-satisfied" people, those personal and group achievements that are pursued by the masses have been irrationally disconnected with the reality of our planet. As such, humans have disrupted natural ecosystem processes through the extraction of natural resources by means of agriculture, mining, unsustainable forest harvest practices, city development, and many other anthropogenic activities. There seems to be an increasing mentality geared towards instant gratification with little thought about the long-term impacts of these actions on natural ecosystems. Such disconnection has resulted in creating novel environmental conditions that outpace the rate at which humans can evolve and as a result we are impacted by several disease conditions such as obesity, myopia, and breast cancer which did not exist during the evolutionary history of our species at the levels observed today.

As mentioned before, our lack of recognition of the cognitive forces that shape the system in which we live, has also placed our species in a fragile situation in terms of valuing natural resources on an everyday basis. In this manner, the air we breathe and the water we drink are expected resources in our daily lives but part of a matrix of externalities in terms of quality and provision. For some social groups it is as simple as "somehow, somewhere, someone is doing something". That level of disconnection, even to recognize that anthropogenic settings are still ecological systems and that forests are only one representation of the latter, has made some groups undervalue the contributions of our species to life on Earth. Even though we cannot deny our predator role in the planet, our species has also been able to interact at a global level with forest resources in all latitudes and biophysical conditions in several other ways.

In such context, the "red queen" race that our species has upheld with the planet and that has affected forest ecosystems in different manners has been well portrayed by the different contributors of the book. Interestingly, as we have read in the different chapters, one aspect in common among the different types of interaction with natural resources has been the exclusion of a framework delineated by responsibility, compromise and respect for future generations. It is perhaps under such scope that nature appropriation should have occurred in the first place. However, as mentioned before, in a historical reality lacking that type of integration with sustainability and with a number of learned lessons, it seems reasonable to think that it is about time to steer the boat in a different direction.

All authors involved in this book, including the editors, wrote this book with one premise in their minds. We did so hoping that the pages of this book could provide elements for personal analysis, growth and integrated elements of discussion toward better practices on our relationship with forest resources and the planet. Even though our relationship with nature, as we have seen in the pages of the book, is deep, long and complex, we hope our work has achieved the goal of setting up different scenarios of human-forest interaction for a better personal understanding of our species position in the network of interactions shaping the existence of life in the planet.

Index